映象新古典
顶级休闲商务会所

ID BOOK 工作室 编

Neo-Classical Reflection
Top Leisure Business Clubs

华中科技大学出版社
http://www.hustp.com
中国·武汉

Modern Club Impression

Club originated from Europe. It is also called clubhouse or chamber. Chamber can also be used as an adjective, meaning "private, interior." From its verbal meaning, we know that club is also a place for people to gather and carry out activities such as entertainment, business and sport.

Since it came into being, club is a place for social elite having a common goal to enjoy their leisure time. Later on, it spreads to every corner of the world, then giving birth to various kinds of schools and patterns. Recently, China's club enjoys rapid development. There are peculiar clubs, multi-functional clubs, boutique clubs, etc., with even more comprehensive functions and higher and higher levels. This developing speed is closely related with the development of the society and the improvement of life quality of the people.

Currently, most private clubs have membership system, with high standard of privacy and short service chain, providing functions such as dieting, meeting, body-building, etc. And such functions are mostly displayed in providing services for daily life and social life for members. And these two aspects are only a little part of the work and life for elite. The future club service would expand towards the breadth and depth of service. With the change of time, this kind of reception clubs is favored by more and more companies. It can be said that the senior commercial clubs would certainly occur in the market-oriented economic system. From social perspective, it is endowed with some business colors and becomes part of business activities. But up to now, clubs are still confined to articles of luxury. This accurately illustrates that club symbolizes quality. I have found that many business services try to prove its service quality through various methods. Thus the business concept of club becomes the main stream of market competition. With approval of more corporations, there would be greater requirements for clubs by people, thus naturally club space design becomes the high-end category of design.

No matter how good the decoration is, it can age in years. It is the people, the service that would make club last long. In foreign countries, club is part of the traditional culture, and some clubs can easily invite world-class elite and renowned scholars. It is out of its cultural ethos, and the collection of culture does not succeed within a short time. This is somewhat related to the cultural customs and habits that senior business club sticks to. Club is connected with the architecture and the surrounding environment. With ecological aesthetic viewpoints, the relationship between human beings and nature, between human beings and the architecture is researched on as an ecological system and organic unity. It is not a research on individual people independent of nature and architecture, nor a research on subjective nature and architecture independent of human beings. Only through harmonious coexistence of human beings and nature can healthy, natural and harmonious human residence be achieved.

Daohe Design Firm
Gao Xiong

当代会所印象

"会所"的前身即"俱乐部",起源于欧洲,英译为"clubhouse",或者"chamber"。"chamber"也可以作为形容词用,意为"私人的、室内的"。单从其字面意思来理解,会所亦是人们聚集一处,进行娱乐、商务、运动等各类活动的场所。

其诞生的最初就是供志同道合的社会名流休闲娱乐的场所,后来逐渐传播到世界的各个角落,进而衍生出各式各样的流派与模式。近几年,国内的会所发展迅猛,独辟蹊径的特色会所、多功能会所、精品会所等,功能越来越全,档次也越来越高。这样的发展速度是与社会的进步和国人生活品质的提升紧密相关的。

目前私人会所多为会员制,私密性相当高,提供的服务链比较短,主要是餐饮、会议、康乐等功能,并且这些功能更多体现在为会员的日常生活和社交生活提供服务,而这两个方面只是高端人士工作生活的一小部分。未来的会所服务将在现有基础上向广度及深度两个方面拓展。随着时代的变迁,这类接待会所被越来越多的企业所青睐。应该说,高级商业会所是在市场经济体制中必然会出现的,从社会角度来说被赋予了一定的商业色彩,成为经营性活动的一部分,不过到目前为止还仅局限于一些奢侈品应用。这也恰恰说明了"会所"就是品质的象征。个人观察发现,时下众多商务服务总是用各种方式来证明自己的服务品质,相应的"会所"的经营理念成为市场竞争的主流趋势。企业的认同越多,人们对"会所"的需求就越大,"会所空间设计"自然成为设计门类中的高端学科。

再好的装潢也会过时,人的因素、服务的因素才是会所能够历久弥新的关键。在国外,会所是其传统文化的一部分,有些会所很容易就能请到世界级精英或者著名学者,就是凭借它的文化气质,而文化的积累不是在短时期内就能完成的,这与高级商业会所秉承的文化传统习惯不无关系。将会所这个主体与建筑及周围环境相结合,加入生态美学的观点,把人与自然、人与建筑的关系作为一个生态系统和有机整体来研究,既不是脱离自然和建筑去研究孤立的人,也不是脱离人去研究纯客观的自然和建筑。只有达到人与自然和谐共生,才能真正实现健康、自然、和谐的人居观。

道和设计机构

高雄

006	悦府会 Yuefuhui Club	120	净雅会所式餐厅未来城店 Jingya Restaurant Club Future City Store
024	枫丹白露悦府会 Fontainebleau Yuefuhui	128	景瑞绍兴望府售楼会所 Jingrui Shaoxing Wangfu Sales Club
044	皇室格调 Royal Tone	136	1982红酒音乐生活馆 Chengdu 1982 Red Wine, Music and Life Club
054	梵克雅堡私人会所 Van Cleef & Arpels Private Club	148	方直君御企业会所 Fangzhi Junyu Corporation Club
064	江滨一号 Riverside No. 1	156	中联天御会所 Zhonglian Tianyu Club
074	逸泉山庄 Yiquan Mountain Villa	166	北海天隆三千海高尔夫会所 Beihai Tianlong Sanqianhai Golf Club
080	贵州铜仁上座会馆 Guizhou Tongren Shangzuo Club	176	十二橡树会所 12 Oaks Club
092	硚口金三角销售中心 Sales Center of Qiaokou Golden Triangle	182	品上·光聚汇 Pin Shang·Guang Ju Hui
098	圆中秘境 Round Secret Fairyland	188	东方银座中心城 Oriental Ginza Central Town
108	常州九龙仓国宾一号会所 Changzhou Jiulongcang Guobin No. 1 Club	198	东方银座会馆 Oriental Ginza Club
		206	阳光金城别墅会所 Sunshine Gold City Villa Club

CONTENTS

222 惠州高尔夫会所
Huizhou Golf Club

228 榕都晶华
Rongdu Jinghua

234 海通一号
Haitong No. 1 Mansion

240 水煮工夫茶道会所
Shuizhu Gongfu Tea Club

246 印象客家
Hakkas Impression

254 三明会所
Sanming Club

260 茗仕汇·茶会所
Ming Shi Hui·Tea Club

268 第一谈紫砂会所
First Talk Dark Red Enameled Club

274 大连亿达庙岭会所
Dalian Yida Miaoling Club

280 西溪·海会所
Xixi·Ocean Club

286 华景里售楼会所
Huajingli Sales Club

292 九乐会所
Jiule Club

298 三省园高尔夫会所
Sansheng Garden Golf Club

306 虎门名店私人会所
Humen Famous Shop Private Club

312 三亚半岛蓝湾接待会所
The Reception Club Of Sanya Peninsula Blue Bay

318 贵族会馆——红酒、雪茄、咖啡吧
Aristocrat Club – Red Wine, Cigar and Coffee Bar

324 碧海国际水疗会所
Bihai International Hydrotherapy Club

338 一湖会所
Yihu Club

Yuefuhui Club

悦府会

设计单位：深圳市昊泽空间设计有限公司	Design Unit: Shenzhen Haoze Space Design Co., Ltd.
设 计 师：韩松	Designer: Han Song
项目地点：浙江省宁波市	Project Location: Ningbo in Zhejiang Province
项目面积：850 m²	Project Area: 850 m²
主要材料：白沙米黄大理石、虎檀尼斯、泰柚	Major Materials: White Sand Beige Marble, Sandalwood, Teak
摄　　影：江河摄影	Photographer: Jianghe Photography

本项目以柏悦酒店为依托，傍依浙江宁波东钱湖自然景区，独享小普陀、南宋石刻群等人文景观资源，地理位置无可比拟。

在空间和视觉语言上与柏悦酒店完美对接。在空间上以中国建筑传统的空间序列强化东方式的礼仪感和尊贵感，在视觉上通过考究的材料和独具匠心的工艺细节，以简约的黑白搭配一气呵成，展现了东钱湖烟雨濛濛、水墨沁染的气韵。

在硬件和智能化体系上坚持柏悦酒店一贯高品质的传承，让客户不经意间感受到骨子里的柏悦性格。比如：一进入会所，所有的窗帘为你徐徐打开，让阳光一寸寸地洒进室内；按一下开关，卫生间的门就会自动藏入墙内；全智能马桶自动感应工作……随处都力求让人感受到高品质的舒适体验。

设置独立专属的高端客户接待空间，如独立酒水吧、独立卫生间。尽享尊贵、专属的接待服务。

细分功能空间，将一个空间的多重功能拆解细分，每个都尽善尽美，大大提升品质感。

增加全新的功能体验，在商业行为中加入文化和艺术气质。我们在地下一层设计了一座小型私人收藏博物馆，涉猎瓷器、玉器、家具、中国现代绘画等……不仅大大提升品质，同时也给客户带来视觉和心理上的全新震撼体验。

身处其中，恍若超脱凡尘，烦恼、杂念消失无踪。带出一抹我独我乐的欢喜。

正所谓：别业居幽处，到来生隐心。
南山当户牖，沣水映园林。
屋覆经冬雪，庭昏未夕阴。
寥寥人镜外，闲坐听春禽。

Neo-Classical Reflection·Top Leisure Business Clubs

This project is set off by Park Hyatt and is beside Dongqian Lake Scenic Spot of Ningbo city, enjoying the human landscape resources of Little Putuo, carved stones from the Southern Song Dynasty, etc. Thus this club has incomparable geological locations.

This club combines perfectly with Park Hyatt in space and visual languages. It strengthens oriental etiquette sensation and noble feelings with traditional Chinese architectural space layers. Visually this club has concise and smooth black and white collocations through exquisite materials and ingenious craft details, displaying misty charms with tradition Chinese painting style of Dongqian Lake.

As for the system of hardware and intellectualization, it sticks to the inheritance of the consistent high quality of Park Hyatt. Thus unconsciously the guests can perceive the essential Park Hyatt characteristics. For example, once you enter the club, all curtains would open slowly for you and sunshine would spread into the room inch by inch. Once you push the button, the door of the washroom would hide into the wall. All-intellectual closestool would carry out auto-induction work... All of these would assure people of high quality comfortable experiences.

This club has independent and exclusive high-

Neo-Classical Reflection·Top Leisure Business Clubs

Neo-Classical Reflection·Top Leisure Business Clubs

end reception room, such as independent wine bar, independent washroom. All guests would enjoy honorable and exclusive reception services.

All functional spaces were subdivided to disassemble the multi-functions of the space. The designer tries to attain the perfect status for every space to promote the quality in a grand scale.

The designer adds brand-new functional experiences, adding cultural and artistic temperament in the business activities. There is a small size library for private collections in the basement, including porcelain, furniture, modern Chinese paintings, jade ware, etc. All these not only uplift the quality, but also create brand-new shocking experiences for the guests visually and psychologically.

It is totally like another world. All worries and distracting thoughts would disappear. People would experience some exclusive joys here.

It is just like the poem goes: Being inside a secluded villa, people would bounce into the idea of becoming a hermit.

The South Mountain is just outside the window and the Fengshui River is echoing the garden landscape.

The roof is covered with winter snow, and the dusk court is shone bright by it.

There are few people coming here, thus we can sit in leisure and enjoy the tweeting of spring birds.

Neo-Classical Reflection·Top Leisure Business Clubs

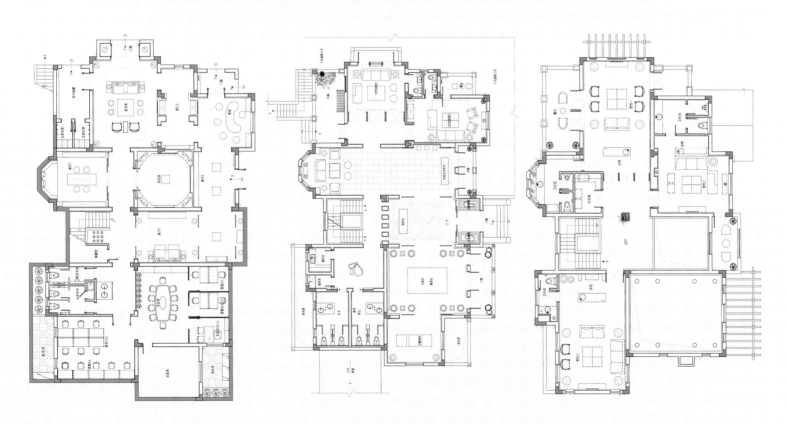

Fontainebleau Yuefuhui
枫丹白露悦府会

设计单位：深圳市昊泽空间设计有限公司	Design Unit: Shenzhen Haoze Space Design Co., Ltd.
设 计 师：韩松	Designer: Han Song
项目地点：浙江省宁波市	Project Location: Ningbo in Zhejiang Province
项目面积：600 m²	Project Area: 600 m²
主要材料：塞维亚米黄大理石、碳化木、雪茄牛皮	Major Materials: Beige Marble, Carbonized Wood, Cigar Cowhide
摄　　影：江河摄影	Photographer: Jianghe Photography

本项目为高端艺术沙龙性质的私人小型会所，提供高端的私人接待及娱乐服务、小型的艺术沙龙展览及高端私人派对。从室内设计到软硬件配置上都提供了顶级、高端的设施及环境。整体空间采用现代英式的木墙板装饰风格，配合上英国 HALO 品牌的家具，以及当代艺术家的艺术作品和有关披头士乐队的纪念物……厚重、粗犷中带着一抹雅皮和时尚。

一、强调生命感。我们希望会所在客人进入的第一步起就感觉生命被激活了一样，灯光由暗变亮，窗帘徐徐开启，让客人感受到最热情的欢迎。

二、在硬件和智能化体系上追求英式的高品质传承。在餐厅和厨房空间中采用分离独立式的中、西厨系统设计，各自配备不同的高端餐厨设备；各空间配备多种灯光场景模式和背景音乐系统；公共廊道采用感应式灯光控制系统；电视机采用隐藏控制系统；红酒吧、雪茄吧有各自独立式的窖藏区；卫生间、衣帽间配有指纹电动感应开启系统；卫浴产品选用特别订制的英国老牌洁具 Christo Lefroy Brooks，彰显贵族风范。

三、强调文化感和艺术气质。由真正的高级管家提供接待服务，为客人讲解设计的细节，每件艺术品背后的故事，每款家具的历史。在这里，可以品尝地道的法国红酒，体验正宗的古巴雪茄。

四、就寝区设计了双主卧系统。并以枫丹、白露各自命题，展现完全不同的视觉效果和居住体验。

五、在服务上与柏悦酒店实现完美对接。如保洁服务、私人酒会等，预约柏悦高级厨师到家中下厨，在地下室设计了专业的 SPA 空间，可预约柏悦 SPA 技师或瑜珈教练到家中提供专属服务。

This project is a private little club with high-end artistic salon characteristics, offering high-end private reception and entertainment services, artistic salon exhibition and high-end private party. From interior design towards soft and hardware configuration, all are supplied with top and high-end instruments and environment. The whole space applies modern English wood wall board decorative style, accompanied with English HALO furniture, artistic works by modern artists, artistic works by modern artists and souvenirs of the Beatles… There is some sense of Yuppie and fashion within this heaviness and wildness.

1. Emphasizing on sense of life. We wish that upon entering the club, the guests would feel like being inspired. The lights turn brighter and the curtain is pulled open slowly. Thus the guests would get the most enthusiastic welcoming.

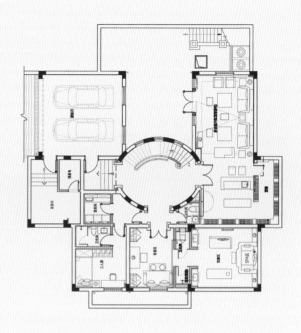

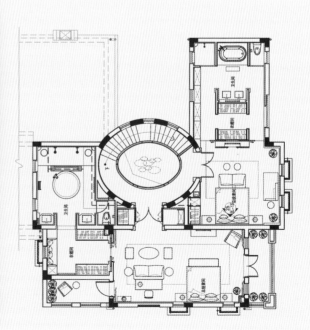

Neo-Classical Reflection·Top Leisure Business Clubs

033

2. For hardware and intellectual system, this project aspires for English high-quality inheritance. The dining hall and the kitchen apply Chinese and western kitchen system design of independent style, allocated with respective high-quality kitchenware equipments. Each space is equipped with multi lighting modes and background music system. The public corridor applies inductive lighting control system. The TV applies hiding control system. Red wine bar and cigar bar have their respective independent cellar storage. Washroom and cloakroom have fingerprinting electro-dynamic operating systems. The washroom uses custom-made British Old Brand sanitary appliances Christo Lefroy Brooks, thus displaying aristocratic temperament.

3. Emphasizing on cultural feel and artistic temperament. This club has real senior housekeeper to provide services, and explain design details, such as the stories behind every artistic object and the history of each piece of furniture. Here, you can taste authentic French wine, and experience the real Cuba cigar.

4. The bedroom area has dual master bedroom system. named Fontaine and Bleau respectively, presenting completely different visual effects and residential experiences.

5. For services, this club is completely connected with Park Hyatt, such as cleaning services, private wine party, etc. You can also book some senior Park Hyatt cook to come here and prepare dishes. There is a professional SPA space in the basement. You can reserve some SPA technician or Yoga coach of Park Hyatt to come here to provide exclusive service.

Neo-Classical Reflection·Top Leisure Business Clubs

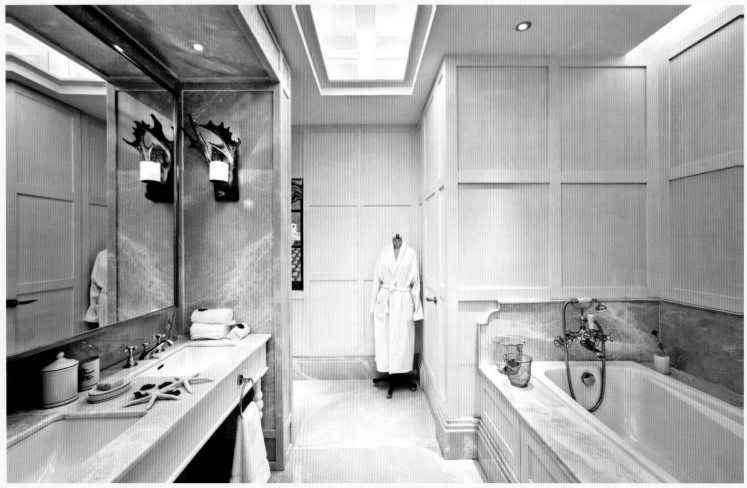

Neo-Classical Reflection·Top Leisure Business Clubs

Royal Tone

皇室格调

设计单位：香港天工设计	Design Unit: Hong Kong Tiangong Design Firm
设 计 师：马治群	Designer: Ma Zhiqun
参与设计：刘思芍、林惠桢	Associate Designers: Liu Sishao, Lin Huizhen
项目地点：福建省福清市	Project Location: Fuqing in Fujian Province
项目面积：800 m²	Project Area: 800 m²
摄 影 师：周跃东	Photographer: Zhou Yuedong

在地产项目规划之初,中恒团队便实地考察了白金汉宫、汉普顿宫、肯辛顿宫等经典的英式皇家建筑,并确定中恒首府联体式住宅和园林景观的英伦情调。为了与中恒首府维多利亚风格的建筑特征相契合,将售楼部的主题定为"皇室格调",旨在将欧洲贵族气息推向极致,不仅诠释英伦风格的复古与奢华,更在空间的雕饰中再现几百年来欧洲艺术文化的缩影。

通过门厅的缓冲进入空间,瞬时让人体会到一种严谨的空间感。浓郁欧洲风情的浮雕壁画夺人眼球,贴上金箔的花线顾盼生辉,大堂穹顶和罗马石柱大胆结合极富趣味,众多元素的搭配营造了很强的仪式感,让人仿佛越过时光隧道,扑面而来的是维多利亚宫廷般的恢弘气势。

在延续手法的运用上匠心独具,通过对变幻多端的局部细节的规划,使得形式上得到视觉的统一。在布局规划上,主要分为模型展示区、洽谈区、英国皇室展示区、咖啡区、3D投影视听室等,在各区域采用开放、宽敞原则,突显区域之间的融合与通透,并与细节融为一体。

地板采用大块面层次分明的大理石拼花,炫目的沙盘坐落于中心位置,上方与之呼应的是大型水晶吊灯与古希腊神话人物的描绘壁画,顶棚壁画则以浓烈的色彩搭配和人物构图巧妙地将人的视线从沙盘引入大堂穹顶。吊顶以壁画为中心,方圆结合的网状辐射向外围延伸,精雕细琢的花线表面被贴上金箔,内镶复古手工油画,显得尤为宏伟壮丽。室内立面的塑造在不失高贵典雅之外顺着场景元素转

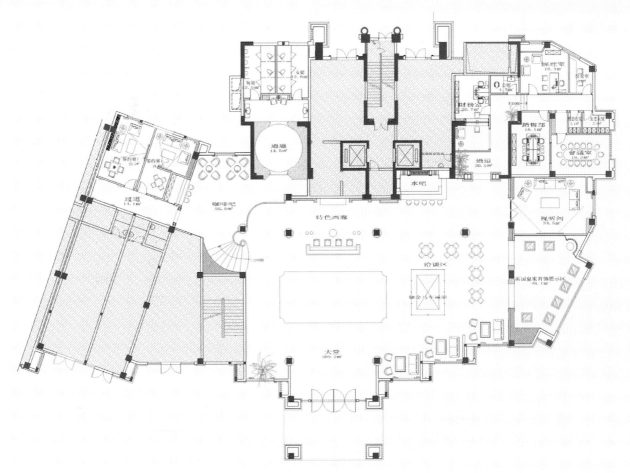

Neo-Classical Reflection·Top Leisure Business Clubs

换变化，使之不显乏味单调。在重复中带着节奏和韵律，宛如正在演奏着一场华丽、高贵、典雅的欧式交响乐。

皇家家具饰品的融合与运用注重比例、色彩、形式的对比协调，尽量缩减家具的比重，保持空间框架本身的美感。在软装的选择上摒弃了柔软的材质和灰暗的色调，主要选用造型感极强的雕刻工艺，迎合顶棚及墙面的细致工艺，在不同形态的统一搭配中，迸发典雅与激情交织的火花。

During the first phase of the project planning, the design team has some on-the-spot investigation towards some English style royal architecture such as Buckingham Palace, Hampton Palace, etc., and defines the English tone for townhouse and garden landscape here. In order to be in accordance with the architecture with Victorian style of this property, the theme of this sales center is oriented to be "Royal Tone," thus to promote the European aristocratic atmosphere to the extreme high level. This not only interprets the restoration and luxury of English style, but also represents the epitome of European art culture for hundreds of years in the decoration of the space.

Once you enter the space from the hallway, you would experience this rigorous space feeling. The relief mural with intensive European charms is very attractive. The colored thread with gold foil shines and the bold combination of the lobby dome and Roman stone pillars is very interesting. The combination of many elements creates some strong ceremonial feel to make people feel like going through the time tunnel and running into this Victorian palace magnificence.

The designer is quite ingenious in the application of approaches. The form attains visual integrity through the planning towards the varying detail parts. As for layout planning, there are model presentation area, negotiation area, English royal presentation area, coffee area, 3D projection room, etc. All areas abide by open and broad principle, and highlight the integration and transparency of all areas, which also integrates with the details.

The ground floor applies large area of marble patterns with obvious division in layers. The shining sand table is located in the central place, corresponding with grand crystal chandeliers and murals of ancient Greek mythological figures. The murals on the ceiling skillfully attract people's eyesight from the sand table towards the dome with its color collocation and figure pictures. The ceiling is centered on the mural. The net combining square and circle extend towards the periphery. The carefully designed colored thread is affixed with gold foil, which appears quite gorgeous with this ancient handmade painting inside. The modeling of the interior facade varies and changes following the scene elements while maintaining nobility and elegance, but not appearing monotonous at all. There are rhythms in the repetition, which makes it seem like a European symphony with nobility, magnificence and elegance.

The integration and application of royal furniture objects focuses on the contrast and coordination of proportion, color and forms. The designer tries to curtail the proportion of furniture and maintains the aesthetic feel of the space structure itself. The selection of soft decoration discards soft materials and dark tones, mainly selecting carving techniques with intensive models and meeting with the delicate techniques of the ceiling and wall. In the integral collocation of different forms, there bursts sparks combining elegance and enthusiasm.

Van Cleef & Arpels Private Club
梵克雅堡私人会所

设计单位：全筑设计	Design Unit: Trendzone Design
设 计 师：陆震峰	Designer: Lu Zhenfeng
项目面积：880 m²	Project Area: 880 m²
主要材料：梦幻金大理石、罗马米黄大理石、紫罗红大理石、金世纪大理石、拼花地板、仿木纹砖	Major Materials: Dream Gold Marble, Roman Beige Marble, Rosso Lepanto Marble, Gold Beige Marble, Block Floor, Wood Grain Tile
摄 影 师：张静	

Neo-Classical Reflection·Top Leisure Business Clubs

影片《绝代艳后》讲述了路易十六的妻子玛丽·安东尼特的生平，她短短的一生也正是洛可可艺术最兴盛的时期。本案的设计灵感来自于洛可可 (rococo) 艺术中独有的浪漫浮华的情调，亮色调、粉色调、奢华的金色或银色的装饰，体现出精制而优雅的装饰主义的浪漫情怀。

本案在构图上也打破了文艺复兴以来的对称模式原则，同时采用色调柔和、高明度、低纯度的粉彩色系，公共区域营造出了古典风格的尊贵与奢华的气息。在材料应用上，设计师选用了多种不同质感的大理石进行装饰。一层以会客、餐饮为主；二层以会议及商务为主；三层为休息就寝区；地下室为酒吧、视听、娱乐、健身、SPA 等功能区。

The film Marie Antoinette records the life of Louis XVI's wife Marie Antoinette. Her short life is the period when Rococo art enjoys its heyday. The design inspiration of this project comes from the romantic and showy emotional appeals exclusive to Rococo art. The decorative objects of bright color tone, pink color, luxurious gold or silver display the exquisite and elegant romantic feelings of decoration.

This project also breaks the symmetry mode principle since renaissance in the aspect of composition of pictures. It also applies soft and bright pink color schemes with low purity. The

public area produces noble and luxurious temperament of classical style. For material selections, the designer selects marble of multiple texture marble as the decoration. The first floor centers on reception and dining. The second floor is mainly space for meeting and business. The third floor is the lounge and bedroom area. The basement contains functional areas such as bar, audio and visual space, entertainment, body-building and SPA, etc.

Riverside No. 1

江滨一号

设计单位：福州林开新室内设计有限公司	Design Unit: Fuzhou Lin Kaixin Interior Design Co., Ltd.
主设计师：林开新	Leading Designer: Lin Kaixin
参与设计：陈晓丹	Associate Designer: Chen Xiaodan
项目地点：福建省福州市	Project Location: Fuzhou in Fujian Province
项目面积：2000 m²	Project Area: 2000 m²
主要材料：大理石、软木板、木纹砖、实木板、老料石材	Major Materials: Marble, Cork Board, Wood Grain Brick, Solid Wood, Stone
摄影师：吴永长	Photographer: Wu Yongchang

传统文化与现代审美的结合,让生活达到了令人愉悦的状态。同样,一个被大众认可的空间,它的优势也往往在于此。江滨一号会所便对这样的空间状态进行了完美地诠释,其透露出来的文化底蕴与艺术气息沁人心脾,并展示着一种自我个性的渗透。

现代钢结构建筑混搭着老宅石阶、廊桥、水景、石臼、石缸、现代漆画艺术、明式家具等,让空灵的视觉享受油然而生。这种繁花落尽的细微之美来自于东方传统文化的思维方式,表面上的"空"实则包含了丰富的精神内涵。于是在二者的矛盾中成就了空间的美感,也体现了江滨一号的精髓。

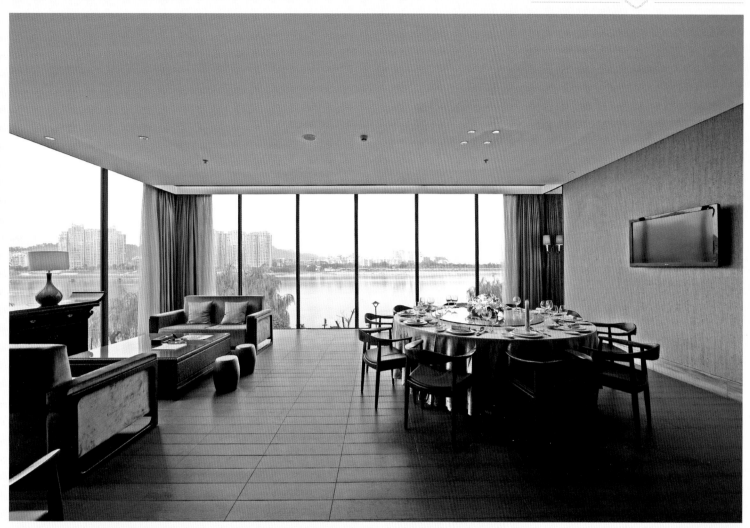

The balance of traditional culture and modern aesthetic values has established a certain pleasing life state. Similarly, the advantages of a widely recognized space lie in this regard. This project is therfect interpretation of such a space state, revealing appealing and refreshing cultural heritage as well as artistic connotations and demonstrating a unique penetration of true personality.

Modern concrete structure mixes with ancient stone steps, bridges, water features, stone mortar, stone jars, modern painting art and Ming-style furniture, creating a natural ethereal visual sense for visitors. This diversified delicate beauty stems from conventional oriental culture's mode of thinking, appearing to be empty but actually containing abundant spiritual connotations. The beauty of the space is therefore achieved from the two contradictory cultural carriers.

Yiquan Mountain Villa

逸泉山庄

设计单位：尚策室内设计顾问（深圳）有限公司
设计师：李奇恩、陈子俊
项目地点：广东省从化市
项目面积：1300 m²
主要材料：黑胡桃木饰面、泰柚饰面、高级手绘壁纸、意大利木纹石

Design Unit: Apex Design Consultants (Shenzhen) Co., Ltd.
Designer: Li Qi'en, Chen Zijun
Project Location: Conghua in Guangdong Province
Project Area: 1300 m²
Major Materials: Black Walnut Veneer, Teak Veneer, High Class Hand-Painted Wallpaper, Italian Wood Grain Stone

本案位于广州逸泉山庄一个人工岛上的独栋建筑，总面积1300m²，四面环湖。整栋建筑经过了重新整合，创造了一个开阔并且开放的生活空间。极简的建筑与其周围的自然景观创造了有机的基调，表现出极大的休闲感。

小区通过一条砾石林荫道到达大院入口，其匍匐的自然情境与高大的建筑形成美观明朗的对比。穿过住宅的入口玻璃玄关可以看到一条由特殊石材连成的"桥"，挑空6m多高的空间，圆弧形的楼梯，创建了非常强烈的视点。一楼有客厅、偏厅、宴会厅、餐厅、厨房及其他服务区，透过宽大的落地玻璃，可以欣赏到岛上的湖景。作为家庭和社交聚会的场所，这里可以容纳不同规模的人群，因此，场地内外的天然过渡显得非常重要。泳池改建在湖面与房子之间，泳池边有一个平台，一直延伸进客厅，一片木甲板成为联系的桥梁。卧室位于二楼及三楼。二楼的西北方向是一个100m²左右的主卧室套房，业主可以在安静、独特的空间里享受生活。面向西北面有一个长10m的露台花园，通过一种简单规则的种植方式表现，以及运用极简手法组合，将室外空间串联起来，形成一条景观连廊。即使在地下一层的挑空庭院区域也融入了柔和的景观，给人一种进入后放松的心情。

湖心岛给人以优雅而闲适的感受，这正是融合了新中式及东南亚元素的体现。室内的家具系列也别具特色，这都是设计的经典作品。灯光及对材质创造性的使用也促成了多样的空间和视觉效果。"设计改变生活"是此案设计师一直坚持的设计之本。

Mid-Lake Island Villa is an independent building located on an artificial island inside Guangzhou Yiquan Mountain Villa, with total area of 1300 m² and surrounded by lake on four sides. Through re-integration, the whole building creates a broad and open living space. Concise style building and its surrounding natural landscape create some organic tone, displaying extreme leisure feel.

We can get to the villa entrance along a boulevard paved with gravel. We get into the garden and pass the stone bridge. Its natural situation produces nice and bright comparison contrasted by high building. Walking past the entrance glass hallway of the building, one can find a bridge linked with special stone material. Other than that, 6 m high space and arc staircase create intensive viewpoint. On the first floor, there are living room, side hall, dining hall, kitchen and other service areas. Through broad French window, one can enjoy the lake view of the island. As a site for family and social parties, this place can accommodate large to small crowd. Thus, the natural transition between the interior and the exterior of the site appears very important. The swimming pool is set between the lake and the house. There is a platform along the swimming pool and extends towards the living room. A piece of wood deck becomes the connecting bridge. The bedrooms are on the second and the third floors. On the north-west side of the second floor is a master bedroom suite which is about 100 m² in size. Thus the property owners can enjoy life in quiet and peculiar space. On the north-west is a terrace garden which is 100 m² in length. Displayed with these simple and orderly planting methods, interior and exterior space is connected together through concise approaches, thus forming a landscape corridor. Even the high courtyard area in the basement also incorporates soft landscape, endowing people with relaxing moods after entering the space.

This island leaves people with elegant and leisurely impression. This is the representation of integrating neo-classical and south-eastern elements. The furniture inside the room has peculiar characteristics. These are all classical design works. Lights and creative application of materials produce varying space and visual effects. "Design can change life" is the principle of design that designer of this case has always been insisting on.

Neo-Classical Reflection·Top Leisure Business Clubs

贵州铜仁上座会馆

设计单位：深圳华空间机构
设 计 师：熊华阳
项目地点：贵州省铜仁市
项目面积：7000 m²
主要材料：大理石、绿可木、布艺沙发、装饰画

Design Unit: Shenzhen Hua Space Institution
Designer: Xiong Huayang
Project Location: Tongren in Guizhou Province
Project Area: 7000 m²
Major Materials: Marble, Lesco Wood, Cloth Sofa, Decorative Picture

坐落于贵州省铜仁市的上座会馆，是接待当地高端人群及高级客商的重要场所。在这样一座有山有水的自然静寂的城中，最适合于它的莫过于传承了几千年历史文化的中式风格建筑。上座会馆由两座3层高的中式建筑构成，从建筑外观到室内设计，再到软装配饰等，均由华空间机构一体化完成。会所内含有健身会所、娱乐酒吧、中式餐饮等项目。

在设计的前期，设计师与甲方深入地沟通此项目的市场定位及各方面要求，并结合当地的消费市场做了调查与研究。根据调研结果，设计师分析出其目标客户的生活喜好、欣赏品位、消费习惯等，并结合当地的经济发展状况及未来的城市发展目标给会馆做长期的发展计划，最终设计出一系列的建筑设计及室内设计方案，世外桃源般的上座会馆由此产生。

设计师在设计项目时采用扬长避短的方式，充分突出了项目的优势。上座会馆依托于当地的山水之景及少数民族的特色，从外观设计到室内设计，采用中式的设计框架结合现代风格的家具、饰品作为装饰。院子中央的小池塘、少数民族特色的壁画、现代风格的沙发、中式古典的木椅、竹叶图案的地毯……使会所由内及外地散发出高雅的新中式设计风格。

空间的设计不在于将墙面及吊顶做复杂的处理，而在于给它恰到好处的点缀。古典中式的沙发是否会有些呆板？设计师结合现代简约的沙发一同陈设，流畅的线条即显现出来；普通的过道如何才能拥有设计感？设计师使用了艺术镜面做墙面，丰富室内层次；正统的中式包房拥有古典的中式家具就足够了吗？我们设计了与众不同的玄关增加了包房的设计感；楼梯的设计是否容易被忽略？设计师注重

项目的每一处细节，在设计中选用了几何形状的时尚楼梯扶手，并在楼道陈列着艺术品……

无论是室内设计，还是产品设计，成功的设计在于相辅相成，由点及面的互相呼应。如此案中多处应用的方圆结合的设计手法，会馆外观、接待大厅、池塘边上、包房内的玄关，都是以方圆为元素的艺术组合。设计师在建筑与池塘之间留有数平方米的空间，并设计成临水而坐的闲叙之地，凉爽的微风、静谧的湖面、舒畅的空间，这一切都受到了客户的喜爱。

对于装饰材料的选用、软装的陈设等，设计师也花费了诸多的心思考察市场。美观且耐用的家具，古典与时尚并存的艺术品，室内装饰材料的颜色搭配，艺术品的摆放角度，绿色植物的装饰，以及使用字体的选择……

总之，项目的设计不仅体现在其"面"上，更重要的体现在其"点"上，在其一点一面的设计与搭配之间，细节的塑造决定着项目的成功。

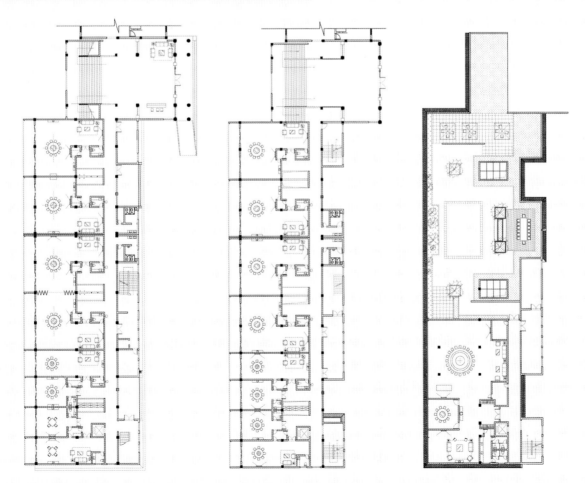

This club is located in Tongren of Guizhou Province and is an important location to receive local high-end customers and senior guests. In this natural and serene town with mountains and water, the most appropriate architecture should be the traditional Chinese style building with several thousand years' history. This club is composed of two 3-storey Chinese style architecture. The building's outlook, its interior design and soft decorations were all completed by Hua Space Institution. Inside this club there are health club, entertainment bar and Chinese dieting, etc.

During the earlier stage, the designer communicated with Party A on the

market orientation and various requirements and carried out investigation and research based on the local consumption market. According to the investigation results, the designer gives out the customers' likes, appreciation tastes and consumption habits, etc. of the end customers. And also the designer carried our long-term development planning based on the local economic development status and the future urban development purpose. Finally, the designer finished a series of architecture and interior design plans. Thus Shangzuo Club which is like a fictitious land of peace came into being.

While designing this project, the designer applies the approach of adopting good points and avoiding shortcomings. The design is rooted in the local landscape and the features of ethnic minorities. From the exterior design to the interior design, it applies the combination of Chinese style design and modern style furniture and ornaments as the decorations. The little pond in the

center of the courtyard, fresco with ethnic minorities' characteristics, modern style sofa, classical Chinese chairs and carpet with bamboo leaf patterns… All these make this club display some elegant new Chinese design style from the interior towards the exterior.

The space design does not lie in making complicated treatment towards the wall and the ceiling, but in making specific ornaments to the point. Does this classical Chinese sofa appear kind of rigid? Combined by sofa of modern concise style, smooth lines thus show up. How can ordinary corridor possess the sense of design? The designer makes use of artistic mirror as the wall surface to enrich the interior layers. Is it enough for the traditional Chinese box rooms to have classical Chinese furniture? We designed some quite uncommon vestibules to add some sense of design for the box rooms. Would the design of staircases be easily neglected? The designer pays attention to every detail and selects fashionable handrails with geometrical shapes. And also there are artistic objects along the corridor…

Be it interior design, or product design, successful design lies in supplementing each other and the points and surfaces echoing each other. Such as the design approach in many aspects, club exterior, reception hall, along the pond and the vestibule inside the box room, all are the artistic combination of elements square and circle. The designer left out several square meters space between the architecture and the pond and designed it into a leisure area along the water. All become the customers' favorite such as cool breeze, quiet lake surface, relaxing space, etc.

As for the selection of decorative materials, soft decoration layout, etc., the designer spent a lot of time investigating the market. Nice and durable furniture, artistic objects combining classical and fashionable style, color collocation of the interior decorative materials, the location of artistic objects, the decoration of green plants and the selection of scripts.

All in all, the design of the project is not only represented in the "surface", but also in the "points". Among this design and collocation of points and surfaces, the shaping of the details determines the success of the project.

Sales Center of Qiaokou Golden Triangle

硚口金三角销售中心

设计单位：尚策室内设计顾问（深圳）有限公司	Design Unit: Apex Design Consultants (Shenzhen) Co., Ltd.
设 计 师：李奇恩、陈子俊	Designers: Li Qi'en, Chen Zijun
项目地点：湖北省武汉市	Project Location: Wuhan in Hubei Province
项目面积：1500 m²	Project Area: 1500 m²
主要材料：玫瑰金、黑檀木饰面、施华洛世奇水晶、所罗门石材、黑金花石材、高级皮纹壁纸	Major Materials: Rose Gold, Ebony Veneer, Swarovski Crystal, Solomon Stone, Black Gold Stone, Advanced Leather Wallpaper

销售中心项目的建筑所在地是武汉市商贸中心地区——汉口镇硚口区，硚口，这颗汉江边上的明珠，正闪耀着夺目的光辉。根据建筑的特殊场地性，设计师将此概念引入设计作品中。

设计师以水为灵感，销售大厅洽谈区上空以水的曲线形态演变而来的水晶灯与椭圆的金钢背景墙及两端的波浪造型墙相呼应，光的透人与景的整合使空间产生了丰富的变化，使之不仅是富有视觉力的饰物，而且也很好地结合了楼盘"星汇云锦"的元素，从而构筑出这个引人冥想的内部空间。

堪比五星级酒店的销售大厅阔绰奢华、金碧辉煌。设计规

划分隔为两层,一层有两间豪华 VIP 贵宾室、挑空 7m 高的椭圆形影音室及六个独立的 VIP 洗手间,还有充满活力时尚的儿童空间,这些传达给来访客户的高贵尊崇的享受。二层为认购签约区及办公区域,空间与众不同的设计调和了现场气氛,达到和谐尊崇的主题。

作为销售中心,某种意义上代表着硚口金三角项目的内在气息,展现出尊贵、时尚的气质,并且彰显着强烈的形象感。

This sales center project is located in the central commercial district of Wuhan – Qiaokou District of Hankou Town. This bright pearl along Hanjiang River – Qiaokou – is shining with dazzling splendors. The designer introduced this concept into the design based on the special regional characteristics of the building.

The designer is inspired by water. The crystal light hanging over the negotiation area of the sales hall evolves from the tracing pattern of water. It echoes with the elliptical background wall and the wavy modeling wall on both sides. The penetration of lights and the integration of views produce rich variations inside the space. This not only is ornament with abundant visual sense, but also combines the elements of the property, thus constructing the inner space arousing meditations.

The sales hall comparable to 5-star hotel is luxurious, resplendent and magnificent. The whole space is divided into two floors. The first floor has two luxurious VIP guest rooms, 7m high oval audio-visual room, 6 independent VIP washrooms and fashionable children's space full of vigor. All these convey aristocratic and noble enjoyments for the visitors. The second floor is contract-signing area and office area. The unique design of the space compromises the site's atmosphere, performing harmonious and aristocratic themes.

As a sales center, it represents the inner atmosphere of this project to some extent, displaying aristocratic and fashionable temperament and intense image.

Round Secret Fairyland

圆中秘境

设计单位：PINKI品伊创意机构＆美国IARI刘卫军设计师事务所
开 发 商：浙江平湖万孚置业有限公司
设 计 师：刘卫军
参与设计：梁义、罗胜文、卢浩
项目地点：浙江省
项目面积：3500 m²

Design Unit: PINKI Innovation Institution & IARI Liu Weijun Design Firm
Developer: Zhejiang Pinghu Vanfu Properties Co., Ltd.
Designer: Liu Weijun
Associate Designers: Laing Yi, Luo Shengwen, Lu Hao
Property Location: Zhejiang Province
Project Area: 3500 m²

Neo-Classical Reflection·Top Leisure Business Clubs

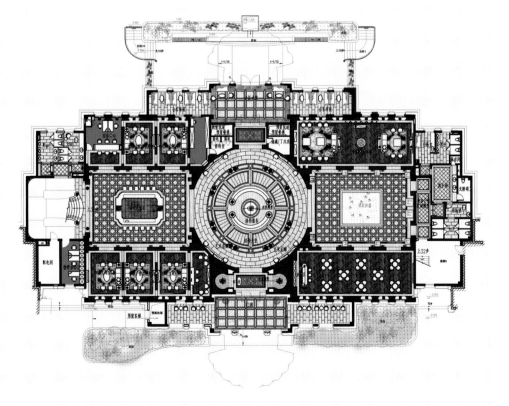

仰望灿烂的星空，追寻内心的法则。欧洲文化的迷人在于人们对生活的向往与自我探索的执着，艺术与哲学、绘画与雕塑、音乐与书籍，无不慰藉和引领人类向往伟大，走向睿智。

本案的立意由此出发，灵动的动线布局带来活泼的空间序列，木与石材等天然材质的运用令空气中弥漫着高贵与舒适，好像徜徉在卢浮宫中聆听毕加索、梵高的心声，又似在维也纳的森林里漫步，亦或是开始了格兰特船长的冒险旅行。

与同类竞争性物业相比，作品独有的设计策划、市场定位。
本项目定位为浙江平湖高端楼盘，面向中高收入客户群。建筑设计方面立意创造经典、可传承来世的作品，因此室内设计注重建筑设计之文化底蕴的渗透与延展，完美诠释项目的尊贵气质，促成整个项目形成一个有机协调的统一体。同时，此项目前期作为销售中心，后期会延续销售中心的空间，在局部改造后作为会所使用。

与同类竞争性物业相比，作品在环境风格上的设计创新点
本案的立意由"圆中秘境"出发，挖掘欧洲经典文化的迷人之处，探索艺术与空间、绘画与雕塑、音乐与生活的关系，糅合国人的生活形态，镶嵌贵族气质的精髓，丰富完备的功能和优雅柔和的气息是其中的亮点。营造出经典、品质、尊贵、可传承、效益并重的意向空间。

与同类竞争性物业相比，作品在空间布局上的设计创新点

在功能布局上遵循了一个重要的原则。前期作为销售中心，做最小的整改后就可以作为会所使用。因此二层设计两个平面方案，一个是销售功能，一个是会所功能，而一层的功能大部分都是可以共用的，这样就保证了设计的延续性，避免了资源的浪费。整个空间讲究轴线对称关系，入口的中庭使一、二层产生了互动与共生，成为空间的亮点，传承了经典与大器。

与同类竞争性物业相比，作品在设计选材上的设计创新点

材料选择上纯净雅致，不管是环保涂料，还是大理石和壁纸。为了烘托出经典和高贵的气质，特意加强了中庭弧形壁的设计，意境深远的油画，呼应着翩翩起舞的舞者雕塑，弧形木质书廊沉淀了几份厚重的历史文脉。没有过多的装饰，极致纯粹的空间弥漫着经典高贵的艺术气息。

You look up at the starry sky and pursue the rules of heart. The fascination of European culture lies in people's longing for life and insistence on personal exploration. The art and philosophy, painting and sculpture, music and books all console people's heart and lead people towards greatness and intelligence.

The concept of this case comes from this. Dynamic space layout creates

active special layers. The application of natural materials such as wood and stone makes the air full of nobility and comfort. It makes people feel like wondering in Louvre listening to the thinking of Picasso and Van Gogh, or walking in the forest of Vienna, or having an adventurous journey with Captain Grant.

In comparison with similar competitive properties, this work has exclusive design planning and market positioning.
This project is oriented to be a high-end property in Pinghu of Zhejiang Province, mainly oriented to medium to high customers. As for the aspect of architectural design, this work intends to become some classical property which can pass through generations. Thus the interior design emphasizes on the infiltration and expansion of cultural connotations of architecture. Thus the noble temperament of this project can be perfectly interpreted, and an organic and harmonious unity can be formed for the total project. At the same time, the earlier phase of this project is used as the sales center. The latter phase is used to continue the space of sales center, and would be used as a club after partial transformation.

In comparison with similar competitive properties, this work has some design innovative points in the environmental style.
The design of this case starts from "secret garden in a circle" and tries to explore the fascinating point of European classical culture. It explores the relationship of

Neo-Classical Reflection·Top Leisure Business Clubs

art and space, painting and sculpture, music and life, and blends the life style of national people. This project is dotted with the essence of aristocratic temperament and is highlighted by complete functions and elegant and soft atmosphere.

In comparison with similar competitive properties, this work has some design innovative points in space layout.
This work follows an important principle in the functional layout. After some minimal alteration, the earlier phase can be used as a club. The second floor is designed with two floor plans: one is the sales function, and the other is the club function. Most of the functions of the first floor can be shared. This guarantees the continuity of the design and avoids the waste of resources. The whole space emphasizes on the symmetrical relationship of axes. The atrium at the entrance creates some interaction between the first and second floor and becomes a highlight in the space, inheriting classicism and magnificence.

In comparison with similar competitive properties, this work has some design innovative points in the selection of materials.
This case selects some pure and elegant materials, be it environmental paintings, or marble and wallpaper. In order to set off classical and noble temperament, the designer highlights the arch-shaped design in the atrium. The painting with artistic conception echoes the dancing sculpture dancers. There is some profound historical context in the arch-shaped wood book corridor. There are no excessive decorations. This pure space is full of classical and aristocratic artistic atmosphere.

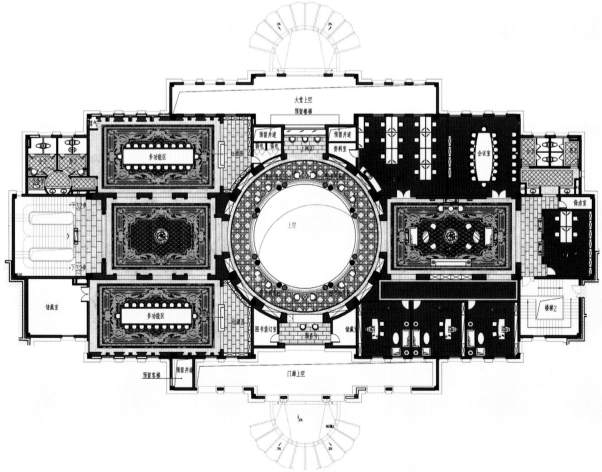

常州九龙仓国宾一号会所

设计单位：上海乐尚装饰设计工程有限公司
设 计 师：卢尚峰
项目地点：江苏省常州市
项目面积：1357 m²
主要材料：大理石、水晶吊灯、陶瓷、青铜、鎏金座钟
摄 影 师：蔡峰

Design Unit: Shanghai Leshang Decoration Design and Engineering Co.,Ltd.
Designer: Lu Shangfeng
Project Location: Chang Zhou in Jiangsu Province
Project Area: 1357 m²
Major Materials: Marble, Crystal Chandelier, Ceramics, Bronze, Gilding Desk Clock
Photographer: Cai Feng

Neo-Classical Reflection·Top Leisure Business Clubs

回望路易十五时期，他带领整个贵族阶层进行了一场轰轰烈烈的浪漫主义革命运动，使法国上层社会在艺术欣赏和生活方式方面，从推崇古希腊、古罗马古典主义转变为追求带东方情调的浪漫主义。室内装饰从顶棚到墙壁，从水晶吊灯到鎏金座钟，玻璃、陶瓷、青铜、大理石材质的各种各样的装饰品、摆件，各式各样的家具，人们不仅要求原料上乘、做工精细，更要求式样新颖、与众不同。

过去时代的家具和装饰品，无论是哥特式的、文艺复兴式的，还是路易十四时期的，统统被弃置不用。设计师和工匠们呕心沥血，绞尽脑汁，以满足这些品位高雅、严格苛刻的王公贵族顾客的需求。正是这一时期，法国产生了对后世影响极大的浪漫主义的"洛可可"装饰风格。

岁月变迁，让浪漫主义在室内设计中有了重新的定义，人们依然崇尚路易十五时期的宫廷气息，但却抛弃了浮夸和臃肿。

本案在设计风格上定位法式新贵族风，在家具造型设计方面，从追求优雅柔和改变为追求宏伟壮丽，家具对称式的布局，更显礼节化，进一步明确空间的商务属性。色彩上大胆运用了红色作为主色调，绿色的遥相呼应让空间形成了一种强烈的戏剧感。不管是窗帘还是花艺、家具面料，都进行一次全新的探索。

19世纪，建筑师德杜克花费数年修复了巴黎圣母院的彩色玻璃，虽然他的努力受到许多批评，但终究经受住了历史的考验，使其作品成为绝世佳品。本案大堂的设计灵感正是基于建筑师的执着，巨大的水晶吊灯缓缓落下，让顶面与地面形成一种空间的联动，使尊客倍感贵族荣耀。

马和钟表主题的设定是让空间的荣耀感有所延续，有战争胜利归来的马，也有休闲出游的马，更有打猎收获的马，只有贵族才能真正拥有这种生活。只有美的极致，才是生活的开始，钟表自诞生之日起就扮演一种非同寻常的角色，光彩夺目背后让我们对历史充满更多的遐想。

VIP男女宾客空间的考虑，在主题上进行了不同定位。男宾客的VIP空间以蓝色作为主色，骑马涉猎展现男人英勇威猛的一面。女宾客的VIP空间以红色、绿色作为空间主色，金色雕花屏风与弗郎索瓦·布歇的《德·蓬帕杜尔夫人》油画的结合，尽显女性风姿。

During the age of Louis XV, he led the aristocratic stratum to have that romantic revolutionary movement in a grand and spectacular scale to revolutionize the super stratum of French society in the aspects of artistic appreciation and life style from advocating ancient Greek and ancient Roman classicism towards romanticism with oriental emotional appeal. As for the interior decoration from ceiling to the wall, such as crystal chandeliers, gilding desk clock, and various decorative objects with materials such as glass, ceramic, bronze marble and various furniture, people not only demand high quality materials and meticulous craftsmanship, but also request its pattern to be novel and uncommon.

Furniture and decorative objects of the past time were all discarded aside, be it Gothic style, Renaissance style, or style of Louis XIV. The designer and craftsmen exert their utmost effort to meet the requirements of aristocratic-like customers with high taste and serious demanding. It is during this period that romantic ROCO Art Deco came into being in France which has enormous influence for the later generations.

Neo-Classical Reflection·Top Leisure Business Clubs

Time goes on, and romanticism has some new definition in interior design. People still uphold the palace atmosphere of Louis XV period, but discarding exaggeration.

The design style of this project is oriented to be French new aristocratic style. In the format design of furniture, it varies from pursuing elegance and softness towards magnificence and splendor. The symmetry layout of the furniture appears much more courteous, further defining the commercial characteristics of the space. The designer selects red as the tone color. It echoes the green color which produces some strong dramatic sensation. The furniture materials display some new exploration from the curtain towards the floriculture.

In the 19th century, a famous architect spent several years to restore the corlorful glass of Notre Dame de Paris. Although his efforts received many critical comments, this work finally stands the test of history and becomes a masterpiece. The design inspiration of the hall comes from the insistance of the architect. This gorgeous crystal chandelier produces some interaction between the ceiling and the ground, All these makes the guests feel very much prioritized.

The themes of the horse and the clock are set to continue the glamour of the space. There are horses coming back from the success of a war, horses having a leisure travel and horses having some hunting. Only people from the aristocratic stratum can really enjoy this kind of life.

VIP male and female guests' rooms have different orientations. The tone color for male space is blye, riding a horse and hunting displaying the heroic side of men. The tone color for female space is red and green, the combination of gold carving screen and the painting Madame de Pompdour by Francois Boucher, displaying female charms to the full.

Neo-Classical Reflection·Top Leisure Business Clubs

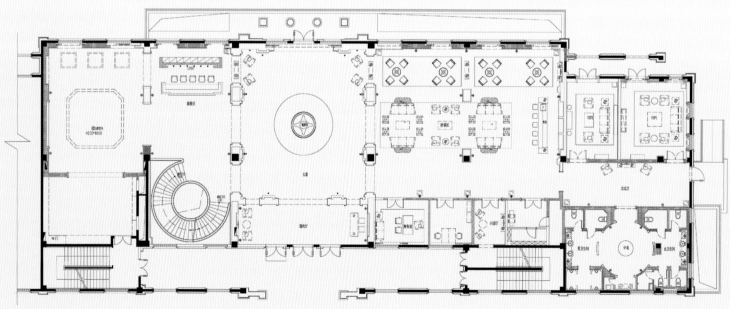

Neo-Classical Reflection·Top Leisure Business Clubs

Jingya Restaurant Club Future City Store

净雅会所式餐厅未来城店

设计单位：睿智汇设计
照明设计：睿智汇设计
设 计 师：王俊钦
项目地点：辽宁省沈阳市
项目面积：10000 m²
主要材料：帝王金石材、香槟金镜面不锈钢、金色柚木防火板、透光云石、水银茶镜、黑色皮革
摄 影 师：孙翔宇

Design Unit: Wisdom Design
Illumination Design: Wisdom Design
Designer: Wang Junqin
Project Location: Shenyang, Liaoning Province
Project Area: 10000 m²
Major Materials: Stone, Stainless Steel, Gold Teak, Fireproof Plate, Marble, Tawny Mirror, Black Leather
Photographer: Sun Xiangyu

美食头等舱——净雅未来城新解读

拥有二十多年餐饮历史的净雅集团，一直以海鲜菜和航海文化而闻名。本案是净雅集团旗下专业提供海鲜美食的餐厅，经营战略定位于商务聚会和私人社交的高端场所。净雅集团在每个餐厅的空间设计上一直寻求突破，有着聘请国际设计师量身打造的发展要求，此次将全新规划的未来城店全权委托于著名设计公司——睿智匯，希望将风格迥异的海洋元素作为主要符号进行诠释，为消费者提供一个私密而极具高雅的视觉就餐体验。

净雅餐厅未来城店位于沈阳市西滨河路，面积为10000m²。是睿智匯团队继多佐日式料理餐厅之后的又一力作，台湾著名设计师王俊钦先生将自然元素手法延续在此案当中，运用后现代设计手法演绎传统文化，强化整体构想，在延续中国传统符号的同时，着重思考地域文化、材料美学、现代科技与当代艺术，创造出一种全新的隐喻古典图像的现代空间。

设计师摒弃了传统高档餐厅的金碧辉煌，在大厅的设计中将"牡丹"、"祥云"、"浮萍"、"瓦片"等元素运用其中，寄托了对东方文化的无限情感。"牡丹"主要表现在顶面造型之中，在恢弘的空间中增添了优雅的气质，烘托清雅闲逸的情境。"祥云"跃然于画面之上，散发着无限曼妙的意境，富有烟云迷蒙的意趣，配合地面水波荡漾的图案，将蜿蜒又悠然自得的神态表现得淋漓尽致，丰富了视觉效果，让人不禁幻想是倘佯在一望无际的大海之中。中庭的设计将柱子与梁的结构大胆地运用和处理，使空间多了份庄严与质朴。传统房屋中"瓦片"元素的运用更加贴近自然。

净雅餐厅未来城店以现代手法重新勾画出私密而高雅的视觉风貌，强化了整体海洋文化特点，让消费者体验尊贵，感受和谐共荣的情怀，在品尝传奇饕餮美食的同时感受到来自净雅集团的文化气息。

First-Class Stateroom of Gourmet—New Interpretation of Jingya Future City Store

Jingya Group has more than 20 years' history in the dieting industry, famous for seafood and navigation culture. This project is a restaurant supplying seafood subordinated to Jingya Group, with operating strategy oriented to high-end location such as commercial gathering and private social activities. Jingya Group has always been searching for breakthroughs for the space design of every restaurant. It has the developing requirements to invite internationally renowned designer to produce custom-made design. And this time, the wholly re-planned Future City store was totally assigned to famous design firm – Wisdom Design, hoping to apply ocean elements as the major symbol for interpretation, and providing private and elegant visual and dieting experiences for customers.

This store is located on West Binhe Rd. of Shenyang city, with area of about 10000 m^2. This is a masterpiece by Wisdom Design team. A famous designer from Taiwan – Wang Junqin – applies natural elements in the design. The designer applies post-modern design approach in the traditional culture, emphasizes on molar construct and focuses on thinking on regional culture, materials aesthetics, modern technology and modern art, thus creating some brand new modern space with the metaphor of classical graphics.

The designer abandons the splendor and magnificence of traditional top-level restaurants and applies elements such as peony, auspicious clouds, duckweed, tile, etc. inside, embodying limitless emotions for oriental culture. Peony is mostly displayed on the ceiling format, adding some elegant temperament for the magnificent space and setting off some elegant and leisurely artistic conception. Auspicious clouds are vividly presented on the painting, sending out some graceful artistic conception and with some interests of mist and clouds. Accommodated with the rippling graphics on the ground, some carefree and content expressions are displayed to

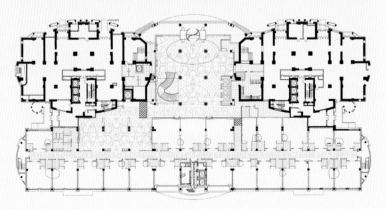

the full, enriching visual effects, while at the same time making people feel like sailing on the ocean. The design of the atrium courageously makes some manipulation and treatment towards the structure of pillars and girders, creating some grandeur and primitiveness for the space. The use of tile elements from traditional houses is very close to nature.

With some modern approaches, this restaurant creates some private and elegant visual atmosphere, strengthening the whole characteristics of ocean culture and making the consumers experience aristocracy and sense emotions of harmony. While enjoying legendary gourmet, the guests can sense the culture atmosphere of Jingya Group.

景瑞绍兴望府售楼会所
Jingrui Shaoxing Wangfu Sales Club

设计单位：上海乐尚装饰设计工程有限公司
设 计 师：张羽
项目面积：936 m²
项目地点：浙江省绍兴市
主要材料：黑檀木、大理石、镀铬金属、琥珀色水晶、彩色皮革、丝绒面料、彩色玻璃
摄 影 师：蔡峰

Design Unit: Shanghai LESTYLE Design and Engineering Co., Ltd.
Designer: Zhang Yu
Project Area: 936 m²
Project Location: Shaoxing in Zhejiang Province
Major Materials: Sandalwood, Marble, Chrome Plating Metal, Amber Crystal, Colorful Leather, Velvet Material, Colorful Glass
Photographer: Cai Feng

本案设计风格定位为：海派 ART DECO，又称"海派装饰主义风格"。整体设计低调、奢华，透露出浓浓的人文趣味。室内设计展现出一种自由、开放的气质，处处显露着海纳百川的胸怀。这就是海派精神，体现了经典与复古的融合。

家具、灯具在材质上运用了具有华丽感的黑檀木、树瘤贴皮、镀铬金属、琥珀色水晶；面料上混合了具有现代时尚感的彩色皮革、丝绒面料；饰品上运用了手工漆画、彩色玻璃等，并且通过拼接抽象的造型加强对材料本身质感的强调，也能赋予老上海以新滋味。

色彩上大的空间区域挑选了纯度比较低的色彩配合空间的文化气质。深紫红、暗灰绿、灰蓝等，与深褐色的木材搭配，处于同一灰度的色彩能增加空间的厚重感，令其充满叙事的味道。而在入口中厅处选择了大胆拼接的法式造型的三人沙发，为空间赋予装饰性，给予视觉新的体验。

装饰画和饰品细节上松针形状的彩色纹路，宝蓝色经典花纹图案玻璃瓶，以及精挑细选的老古董装饰瓶，都让人感觉到海派气息扑面而来。

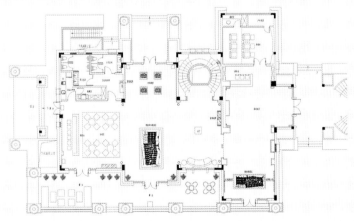
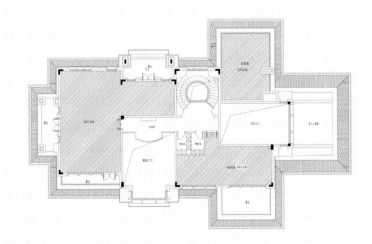

The design style of this case is oriented to be Shanghai style ART DECO. The whole design is low-key and luxurious, revealing intensive human interests. The interior design displays some free and open temperament, conveying the artistic conception of "The sea refuses no river." This is the spirit of Shanghai style, displaying the integration of classicism and restoration of ancient ways.

As for materials, furniture and lighting accessories apply luxurious black sandalwood, lagging, chroming metal, crystal, etc. As for fabrics, there is the combination of colorful leather, velvet materials, etc. There are objects such as handmade lacquer painting and colorful glass. The design highlights the texture of the materials through collage of abstract formats. This can also entrust some new meaning for the old Shanghai.

As for colors, grand space areas select colors with lower purity, accompanied with cultural temperament of the space. The combination of dark purple, dark grey green, grey blue colors with dark brown timbers makes the space appear much more profound, and makes it full of narrative taste. At the entrance, the designer selects French style long sofa to this decorative space give people some new visual experiences.

Decorative painting, colorful lines with pine needle shapes, dark blue glass bottles with classical floral design and carefully selected antique vases give people abundant Shanghai style amorous feelings.

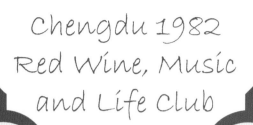

1982红酒音乐生活馆

Chengdu 1982 Red Wine, Music and Life Club

设计单位：深圳市新冶组设计顾问有限公司
设计师：陈武
项目地点：四川省成都市
项目面积：796 m²
主要材料：做旧橡木地板、镂空雕刻板、做旧铜板、香槟金银箔、做旧红砖、金属网帘

Design Unit: Shenzhen NewEra Design Consultant Co., Ltd.
Designer: Chen Wu
Project Location: Chengdu in Sichuan Province
Project Area: 796 m²
Major Materials: Archaic Oak Flooring, Hollow-out Carving Plate, Archaic Copper Plate, Champagne Gold and Silver Foil, Archaic Red Brick, Metal Curtain

本案为坐落于"一座让你来了不想走的城市"——成都市。

西南、暮阳、柏油路与斑马线，荔枝面麻石与黑色烤漆混合元素建筑物伫立、笃定，令人难辨身处亚洲亦或欧洲大陆城市，一抹昏黄怀旧的光源从建筑物泄露，仿佛是上帝的手点燃了武侯大道的马路之光。建筑侧墙体镶嵌"1982"，品牌LOGO避开正面的张扬，出现在东侧面，造型大胆、线条简洁，突显设计师"与晨曦第一时间亲密相见"的设计意图和祈愿品牌的深刻寓意，建筑物整体静穆之境为设计意图的植入提供了圆满的空间围合与精神尺度，蹒跚于光的隧道，悄然点击"1982"，设计师将手笔由建筑物外墙的优雅性有力地植入"1982"的每个角落，绽放后现代的欧陆情调，布局脱俗，强调与众不同的精神轨迹。

"1982"，是集红酒、私人派对、艺术秀、管家服务体系化的生活品鉴馆。伫立大门，一眼望去是3D式线条空间，干净又不失

丰腴，香槟金梁柱驳接顶棚与复式楼地板，左右梁柱盘旋至顶棚平铺成"O"字形修饰，边缘相临，形成视觉效果"8"字形的图饰俘获所有的视觉关注，设计师将品牌字符进行巧妙的空间修辞和植入，从第一眼开始给人注入"简约却不简单"的品牌精神，左右梁柱分别由上而下呈X曲线空间中落幕，从设计出发，依托视觉影响力，植入"1982"的品牌之手，左右摊开，展现礼仪，恭候贵宾，衬托顾客群体的定位与消费品质。从作旧橡木地板、错落有致的木质红酒桶、香槟色金属鸟笼、墙射灯饰的设计选材运用，无一不与质朴、简约、高私密的品牌内涵有着强烈的心理暗示。

走进转角，可见作旧木箱堆叠，以及皮质复古手提箱、红酒瓶与奶白花束、木质老相框，纯白蜡烛伫立于作旧的简约铁质烛台上，不规则木质拼组的静态马饰物，独立体的空间交融，多国元素的融合，设计手法由单纯的装饰升华至与红酒文化所提倡的"无国界艺术装置"，透露品牌专注"圈子"文化的确立。墙面凹陷木质镂空字体设计与灯光聚焦折射"1982"的方向运用，设计师在有限的空间内顺从设计之道充分彰显对于设计、艺术、商业、阅历、生活的驾驭，从细节布局表达品牌对于宾客权限的细致服务，全城唯一一家管家服务生活馆，展现出无处隐匿的谦虚内敛的雅致情怀。

作旧红砖、金属网帘、圆形石桌、皮质围合靠背沙发、裸露石砖顶棚，舞台灯如有白银光芒的保鲜膜覆盖，作旧香槟金与阵列式作旧橡木地板形成由下而上统一格调的古典奢华简约风。在这里，私人home party、美女红酒大使、法国艺术疯马秀，在酝酿昏黄的灯光格局中，行为艺术关怀舒缓平日的疲惫与倦怠。生活的动力，不是一味灌输能量，而应张弛有度，这里不仅有私有化的心理空间，也让人从视觉、味蕾品牌及贴身管家服务中享受到美的事物。心灵SPA，从魅力、亲切、传承、品质开始蔓延，期待来自无界的亲睐。

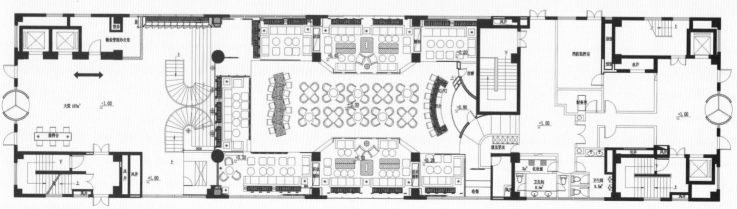

Neo-Classical Reflection·Top Leisure Business Clubs

This project is located in 1F, Elite Club, No. 445, Shuangnanduan Hongpailou Rd., Wuhou Boulevard in Chengdu City which is a city where if you come, you would never want to go.

Southwest, setting sun, asphalt road, zebra crossing, stones and the architecture with some mixed elements such as black stoving varnish… You can never distinguish if you are standing in some place in Asia or a city in the European continent. Some dim and nostalgic light source is revealed from the architecture. It is like the God has ignited the light of road on Wuhou Boulevard. The side wall of the architecture is inlayed with "1982"which logo avoids direct exposure but appears on the east side with bold modeling and concise lines, highlighting the designer's design concept to "have intimate contact with first rays of the morning sun" and profound connotations for this brand. The solemn and quiet state of the architecture provides some perfect space enclosure and spiritual scale for the implantation of the design concept. Tottering in the tunnel of light, quietly clicking "1982", the designer embedded elegance into every corner of this brand name, modern continental appeals bursting into full bloom. The uncommon layout focuses on the unique spiritual traits.

1982 is a life club which systematically integrates red wine, private garden, art show and housekeeper service. Standing at the

door, what comes to the eyes is 3D style space of lines which is clear but well-developed as well. Champagne gold column posts connect the ceiling and the duplex flooring. The beam posts on both sides spiral towards the ceiling, which forms "0". Totem

with visual effect of "8" captures all the visual concentrations. The designer carries out some ingenious treatment towards the brand name and gives people the first impression of "concise but not simple." The stream posts present themselves as "X" streamlines. Starting from design, relying on visual effect, embedded with the brand of "1982", displaying etiquette, welcoming the guests... The whole design sets of the orientation and consumption quality of the customers. As for the archaic oak wood, well-proportioned wood red wine barrels, champagne-colored metal bird cage and the design materials selection of decorative lighting, all have some intensive visual and mental hints for the primitive, concise and private brand connotations.

Around the corner, you can find

piles of archaic wood cases, archaic leather suitcase, red wine bottle, milky white bouquet and wood old photo frame. The pure white candle stands on the archaic concise iron candlestick. There are wood decoration of a horse, space integration of independent objects and multinational elements. The design approach improves from pure decoration towards "borderless artistic objects" that red wine culture advocates for. All these display that this brand focuses on the establishment of "circle" culture. The design of the hollow-out wood font on the wall surface and the focus of lights reflect the orientation of "1982". Within limited space, the designer follows the philosophy of design and displays his mastering of design, art, business, experience and life. And this expresses the meticulous service of this design towards the rights of guests. This is the only life club with housekeeper service and conveys the modest elegant emotions everywhere.

The archaic red brick, metal curtain, round stone table, leather sofa, bare stone ceiling and the stage lights shining with silver color... The archaic champagne gold color and oak flooring have integral color tone and produce classical and luxurious concise style. Here, among the dim yellow lights, performance art can liberate people from the exhaustion and tiredness of daily life. The motive of life does not like in instilling the energy but in flexible degree. There is not only private psychological space here, but also nice things which people can experience from visual sense, taste bud and personal housekeeper service. Mental SPA starts from charm, intimacy, inheritance and quality and aspires for borderless sphere.

方直君御企业会所

设计单位：戴勇室内设计师事务所
软装工程／艺术品：戴勇室内设计师事务所 & 深圳市卡萨艺术品有限公司
开 发 商：惠州方直实业有限公司
项目地点：广东省惠州市
项目面积：1000 m²
主要材料：木纹米黄云石、黑檀木纹云石、灰麻石、真丝手绘壁纸

Design Unit: Eric Tai Design Co., Ltd.
Soft Decoration and artistic Object Designer: Eric Tai Design Co., Ltd. & Shenzhen Kasa Artistic Objects Co., Ltd.
Developer: Huizhou Fangzhi Industrial Co., Ltd.
Project Location: Huizhou in Guangdong Province
Project Area: 1000 m²
Major Materials: Wood Grain Beige Marble, Black Ebony Wood Grain Marble, Grey Stone, Real Silk Hand-Painted Wallpaper

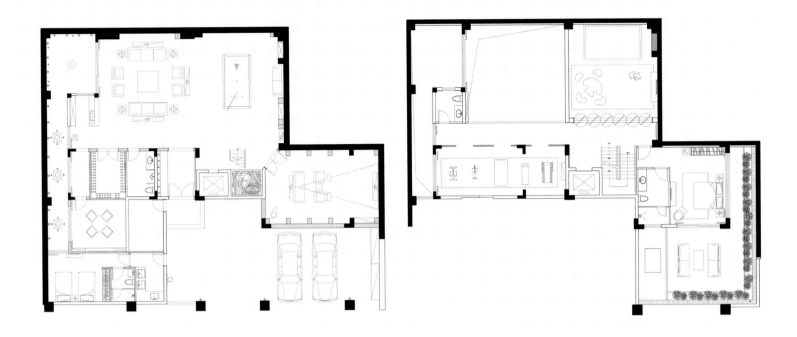

东方人在很久以前就发现了枯槁的美感，在深山古寺、暮鼓晨钟、枯木寒鸦、荒山瘦水中，追求一种独特的韵味。这也是与大巧若拙哲学相关的境界。东方艺术的最高境界是写意，追求挥洒自如、一气呵成和苍劲质朴的艺术造诣。在本案中，设计师意图达到同样的境界，传达大气、质朴及优雅的空间效果。

把东方传统的气韵蕴藏在宽阔、开敞的大尺度的空间中，通过中国传统建筑中轴对称的平面布局和立面构图、中国印象的黑白色调、巨幅的线描中国山水、高耸到顶的中式隔屏，以及精心设计的原创中式家具，呈现出现代中式空间的大气。

室内处处都摆放着设计师精心设计的陈设品，如家具、灯饰和挂件，也陈列着种类繁多的中国物件，旧家具、陶瓷、线装书、黑檀四合院摆件、佛像、茶具、文房四宝、画轴，这些古老的物件与现代的设计在恰当的空间里邂逅，让现代设计蕴涵着深深的历史印迹，让古老的中国艺术焕发活力。设计师对中国元素准确的把握、运用和组合，在传达尊贵大气的同时，让人切实地感受到东方的人文古韵。

Long time ago, oriental people have found the aesthetic beauty of "withered stuff". They have some peculiar charms in remote mountains, ancient temples, evening drum and morning bell in a monastery, dead wood and jackdaw, barren mountain and sheer water. This is related to the artistic concept of "great wisdom appearing stupid" philosophy. The tidemark for oriental art is freehand brushwork, acquiring some spontaneous, coherent and primitive artistic attainments. For this case, the designer intends to attain same quality, thus conveying magnificent, primitive and elegant space effects.

The designer tries to contain oriental traditional charms in the space with broad and large scale, displaying the magnificence of modern Chinese space through axis symmetric plan layout and façade composition in traditional Chinese architecture, impression black and white tone of China, large scale line drawing Chinese landscape, high Chinese partition and carefully designed original Chinese furniture.

There are furnishings carefully designed by the designer everywhere in the room, such as furniture, lighting accessories and pendants, along with a wide range of traditional Chinese objects, such as old furniture, ceramics, thread-bound books, ebony objects from quadrangle dwellings, figure of Buddha, tea set, writing materials, painted roll, etc. These old objects meet by chance with modern design, thus endowing modern design with profound historical imprints, finally making traditional Chinese art glow with vitality again. The designer masters traditional Chinese elements well and applies them in a proper way. All these make people feel oriental human charms in a real way while conveying aristocratic and magnificent atmosphere.

Zhonglian Tianyu Club
中联天御会所

设计单位：福建国广一叶建筑装饰设计工程有限公司
设 计 师：何华武
方案审定：叶斌
项目地点：福建省福清市
项目面积：4500 m²
主要材料：帝皇金大理石、黑金花大理石、西班牙米黄大理石、黑檀木、皮革、镜面、不锈钢、贝母陶瓷锦砖

Design Unit: Fujian Guoguang Yiye Architectural Decoration Design and Engineering Co., Ltd.
Designer: He Huawu
Project Examiner: Ye Bin
Project Location: Fuqing in Fujian Province
Project Area: 4500 m²
Major Materials: Stone, Black Gold Marble, Cream-Colored Marble, Leather, Mirror Surface, Stainless Steel, Mosaic Tile

建筑，常常被誉为"凝固的音乐"，因为它和音乐一样给人以美好的感受和体验，会所的整体设计风格，从气质上发扬了欧式设计文化的精髓，强调了室内空间与自然的相互交流。古往今来，从东方到西方，人们一直追求着居住空间的最高境界，与环境的融合。

空间艺术的生命在于不同功能空间的有效组合，从而形成独特的空间效果，让不同功能空间通过有效的组合，产生令人难忘的艺术效果。设计师采用独特的设计手法将不同功能的空间有机整合，成为天御会所倡导的新生活方式。室外咖啡区域中，室内空间与室外环境的和谐融合，给人以莫大的惊喜，形成戏剧般的高潮，让人独享一份宁静与闲适生活韵味。

设计的生命在于传承和创新，正如建筑设计史上一代代大师一样，一方面需从普通性的原则入手，另一方面不断追寻自我个性的体现，通过风格化的空间设计表达设计的思想，最终达到臻于完美的境界。

Architecture is usually honored as "solidified music". As just like music, it can give people nice feelings and experiences. The whole design style of this club promotes the essence of European design culture from the aspect of temperament. It emphasizes on the mutual communication between interior space and nature. From ancient times till now, from east to west, people have always been pursuing the ultimate realm of residential space, i.e. the integration with the environment.

The life of space art lies in the effective combination between different

Neo-Classical Reflection·Top Leisure Business Clubs

functional spaces, thus to form peculiar space effects. The valid combination of different functional spaces can produce unforgettable artistic effects. The organic integration of different functional spaces through peculiar design approaches becomes the new lifestyle that Tianyu Club calls for. For the exterior coffee area, the harmonious integration of interior space and the exterior environment gives people some grand surprise. This dramatic design lets people enjoy some tranquil and relaxed life style.

The life of design lies in inheritance and innovation, just like generations of masters in the architecture history. On one side, we need to start from the ordinary principle. On the other side, we need to pursue the presentation of personal characteristics. The design displays the concepts of design through stylized space design, ultimately attaining perfection.

Neo-Classical Reflection·Top Leisure Business Clubs

Beihai Tianlong Sanqianhai Golf Club
北海天隆三千海高尔夫会所

设计单位：深圳大易室内设计有限公司
设 计 师：邱春瑞
开 发 商：北海天隆房地产开发有限公司
项目地点：广西省北海市
项目面积：9600 m²
主要材料：海浪灰、意大利木纹、古木纹、鱼肚白、黑伦金、雨林绿、亚洲米黄、金香玉、黑金砂等大理石

Design Unit: Shenzhen Dayi Interior Design Co., Design
Designer: Qiu Chunrui
Developer: Beihai Tianlong Properties Development Co., Ltd.
Project Location: Beihai in Guangxi Province
Project Area: 9600 m²
Major Materials: Grey Marble, Italian Grey Grain, Ancient Wood Grain, White Marble, Black Gold Marble

本案以现代欧式风格为主体，高档的材质配合流畅的线条，将世人所倾慕的欧风韵完美演绎出来。新中式元素和古典欧式元素的适当穿插点缀，加强了整体空间的设计感，并体现其对不同风格的包容，满足了不同会员的需求。墙壁和地面所采用的装饰材料丰富多样，根据不同空间的表现需要而定，石材、木材、地毯一应俱全，图案呈多元化形式，丰富了空间的视觉内容。在软装饰上，设计师选用了大量考究、高档的物品，以少而精的方式呈现于空间中，这样既避免了过多烦琐的东西导致人眼花缭乱，也体现出空间本身的高品位和不拘一格的特性。此外，会所大部分的墙面都用石材来装饰，石材上的天然纹理，有的如大山般尽显磅礴之气，有的与室内的各种线条遥相呼应，大大提升了室内的空间感，让整个会所显得奢华大气。

The main style of this project is European style. The combination of top materials and fluent lines produces perfectly the elegant European style that people have an admiration for. The alternation and ornaments of new Chinese elements and classical European elements strengthens the design of the whole space, displays the inclusiveness of different styles and meets the requirements of different members. The decorative materials of walls and floors are varied and are decided by the presentation requirements of different spaces. There is a whole list of stones, woods and carpets and the diversified patterns enrich the visual perception of the space. As for the soft decoration, the designer selects quite a number of high-end objects and displays them in the space with smaller quantity but better quality which not only avoids much too complicated things dazzling people's eyes but also displays the high taste of the space and its peculiarities. Other than that, most of the walls of the space are decorated with stones with natural texture, while some are like majestic mountain, and some echoes various lines inside the room which greatly uplifts spaciousness of the space and makes the whole club appear luxurious and magnificent.

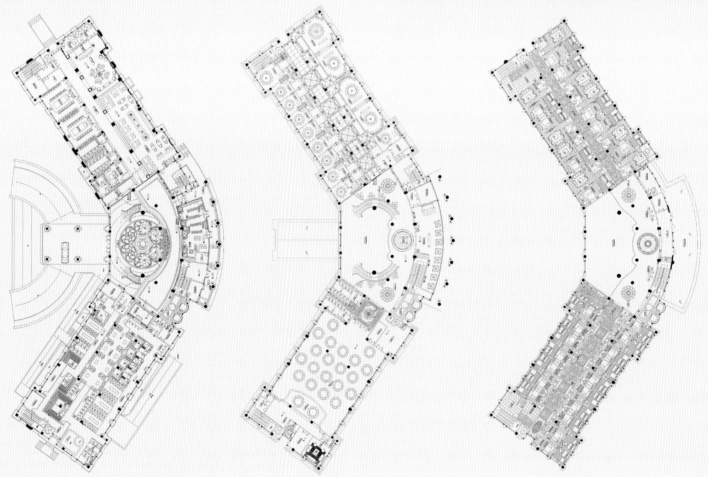

Neo-Classical Reflection·Top Leisure Business Clubs

12 Oaks Club
十二橡树会所

设计单位：PINKI 品伊创意集团 & 美国 IARI 刘卫军设计师事务所
配饰设计：品伊创意机构 & 知本家陈设艺术机构
设 计 师：刘卫军
项目地点：广东省深圳市
设计面积：2000 m²
主要材料：沙漠金大理石、阿曼米黄大理石、深啡网大理石、贝壳陶瓷锦砖、金箔、紫芯影饰面板、地毯

Design Unit: PINKI Innovation Institution & IARI Liu Weijun Design Firm
Decoration Design: PINKI Innovation Institution: Zhibenjia Layout Art Institution
Designer: Liu Weijun
Project Location: Shenzhen in Guangdong Province
Project Area: 2000 m²
Major Materials: Imported Granite Marble, Oman Beige Marble, Dark Coffee Net Marble, Shell Mosaic Tile, Gold Foil, Veneer, Carpet

Neo-Classical Reflection·Top Leisure Business Clubs

在玛格丽特·米切尔笔下，十二橡树庄园是美好、浪漫的代名词。

在十二橡树庄园，有自然的旖旎风光和豪华的酒宴、风趣的野餐、迷人的舞会、漂亮的服装、风度翩翩的绅士和青春俏丽的名媛……正如《飘》所说，它"豪华而骄傲，象征一个阶层和一种生活方式"。十二橡树庄园，承载着对美好生活的向往和憧憬，蕴涵着对浪漫爱情的守护与坚持，寓示着对温馨家庭的守望和眷恋。这一充满了历史情趣的美丽地方，19世纪美国南方如诗如画的庄园生活，如今，历经数百年的时光洗练，在深圳再次展现她的曼妙身姿。

十二橡树会所为深圳唯一原石山别墅、230 m礼宾大道，纯智能化小区，双拼、叠拼及小高层平面房，经典的北美庄园范本，引领现代别墅生活新潮流。深圳十二橡树会所是北美风格在中国的发展与提炼的成果，注重建筑细节、具有古典情怀、外观简洁且大方。

十二橡树会所为2层北美建筑风格，一层主要为接待区、楼盘展示区、奢华体验区、休闲洽谈区等，运用石材浮雕、大理石拼花、水晶吊灯、大型壁画等展示着整个项目的恢弘大气。通往洽谈区的走廊，一幅幅逼真的壁画，仿佛重现19世纪奢华的贵族生活。二层有VIP室、洽谈室、露台及办公区等。园林风格以新古典主义与自然主义相结合，运用意大利造园手法及托斯卡纳的生活情调，整体氛围体现自然、生态的风情。

In the eye of Margaret Mitchell, 12 Oaks Manor represents beauty and romance.

There are nice landscape, luxurious feast, picnic which is full of fun, enchanting dancing party, nice clothes, graceful gentlemen and pretty ladies… Just like what is stated in Gone with the Wind, this manor is "luxurious and proud, representing a class and some lifestyle." This manor embodies people's longing for the happy life, contains people's insistence on romantic love and implies the attachment to a warm family. This is a nice place with historical interests which provides some picturesque manor life in the south of America in the 19th century. Now, after hundreds of years, this manor redisplayed her beautiful figure in Shenzhen of China.

This is the only villa with original rocky mountain in Shenzhen. This is a pure intelligent community with 230m long avenue, binary family house, duplex family house and 7~11-floor building. This is a classic north-American manor example and would lead the new trend of modern villa life. This project is the development and improvement of North-American style in China. It focuses on architectural details, possesses classical allusions and has concise and generous outlook.

This club is two-storey North American architectural style. The first floor is mainly reception area, presentation area for properties, luxury experience area, leisure and negotiation area, etc. The whole project appears very magnificent with stone sculpture, marble pattern, crystal chandeliers and grand mural, etc. The corridor leading to the negotiation area is decorated with many murals. It seems as if 19th century aristocratic life reappeared.

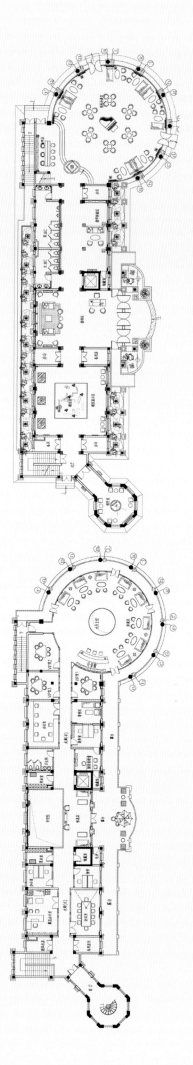

Pin Shang · Guang Ju Hui

品上·光聚汇

设计单位：道和设计机构
项目面积：180 m²
主要材料：乳化玻璃、红橡木饰面板、深灰色地塑、雅士白大理石、灰镜、白色烤漆玻璃
撰　　文：Jenny
摄 影 师：李玲玉

Design Unit: Daohe Design Firm
Project Area: 180 m²
Major Materials: Emulsified Glass, Red Oak Wood Veneer, Dark Grey Plastic Flooring, Jazz White Marble, Grey Mirror, White Baking Varnish Glass
Composer: Jenny
Photographer: Li Lingyu

镜中的安逸世界

这是一个以镜面为核心元素打造的舒适的茶空间，设计师将镜面进行合理的布置，通过其反射与投影，使得空间有着一种神秘之感，让人有一种一探镜中世界的强烈愿望。穿过用镜面布置的走廊，看着空间的镜中倒映着周遭的场景，分不清什么是真实的，什么又是虚幻的。镜子的反射作用使得空间少了些压抑，看起来更加大气磅礴。

在射灯所散发出的柔光中来到主厅，这个功能区域的布置十分素雅，白色的墙壁搭配上深灰色地塑地板，墙壁则用磨砂玻璃作为装饰。中部空间放置着木质的长桌，配上中式圈椅，古朴、自然的木色流露出温馨的气息。长桌穿过用一个磨砂玻璃做成的中式花窗，在花窗的内部埋设射灯，当灯被打开时，灯光便会透过磨砂玻璃散发出来，十分富有巧思。

空间的另一侧则摆放着略显现代的黑色皮质沙发套件，皮质有着特色的纹路，舒适感十足。而黑色大理石纹理的茶几，则带着几分深沉，一旁的白色球椅给这略带沉闷的色调带来一丝活力，缓和了空间的气场。灯光与玻璃是这个空间的主题，它们在这个空间内互相交融，给空间创造出另一番风味，而随处可见的特色雕塑也散发着艺术的气息。

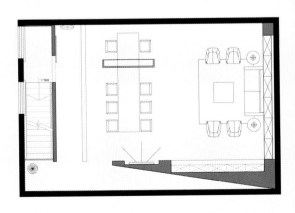

Serene World in the Mirror
This is a comfortable tea space created with mirror surface as the core element. The designer makes some appropriate treatment towards the mirror surface and creates some mysterious feeling inside the space with reflection and projection, making people have the intensive wish to explore the world inside the mirror. Through the corridor decorated with mirror, the mirror reflects everything around. People cannot recognize what is real, what is false. The reflection of the mirror makes the space appear not that depressing, but more magnificent.

People come to the major hall in the soft light created by the spotlights. The arrangement of the functional area is very elegant. White wall is collocated with dark grey plastic flooring. The wall is decorated with frosted glass. In the middle area there is the wooden long table, accompanied with Chinese round-backed armchair. The primitive and natural wood color sends out some warm atmosphere. The long table runs through a Chinese latticed window made of frosted glass. Inside the latticed window there are spotlights. And when the lights are turned on, light would send out from the frosted window, quite ingenious.

On the other side of the space, there are black leather sofa accessories with some modern feel. The leather has some peculiar pattern, with strong sense of comfort. The tea table with black marble pattern is kind of profound. The white chair aside brings some dynamic feeling for the dumb colors, softening the atmosphere of the space. Lights and glass are the theme of the space, integrating with each other in the space and creating some other aroma for the space. Other than that, peculiar sculptures are everywhere in the space, with enormous artistic atmo.

Oriental Ginza Central Town

东方银座中心城

设计单位：深圳市盘石室内设计有限公司
设 计 师：吴文粒
项目地点：辽宁省盘锦市
项目面积：3500 m²
主要材料：维也纳米黄大理石、皮革、玫瑰金

Design Unit: Shenzhen Panshi Interior Design Co., Ltd.
Designer: Wu Wenli
Project Location: Panjin in Liaoning Province
Project Area: 3500 m²
Major Materials: Vienna Beige Marble, Leather, Rose Gold Stainless Steel

位于辽宁盘锦的东方银座中心城会所，周边环境优雅宁静，简约淡雅的整体色调极具东方意味，它所营造出的是一种尊贵、私密的空间氛围。一进入室内，就会被它磅礴的气势所吸引，高挑的顶棚设计夸大了空间的尺度感，而简洁的吊顶强调的是空间的秩序感。现代时尚感与东方元素的完美融合深植于整个空间中，蕴含了大气深邃的东方意境，并符合当下的审美趣味。

设计从东方传统元素中汲取灵感，大胆地加以"破坏"和"否定"，从而创造出一个全新的设计理念。设计师善于在传

统符号中寻找能够抽象表达的设计元素，借以触碰心灵中最感性的地带。落地窗前设计师别具匠心地运用中式隔扇进行遮挡，阳光透过缝隙照射进来，使得隔扇熠熠生辉，带动起空间的活力。就座区使用圆形进行围合，不仅形式感强烈，也能让人在其中感受到空间环境相互交融的和谐感。在文化、艺术的表达形式背后，设计师更愿意肆意地表达出一种艺术的力量，这在无形中增强了空间的神秘张力。

This club is located inside Oriental Ginza Central Town in Panjin of Liaoning Province. This is a quiet and elegant location. The concise, simple and elegant overall color tone displays intensive oriental connotations. What is produces is some aristocratic and private space atmosphere. Once you enter the room, you would be attracted by its magnificent atmosphere. The lofty ceiling design exaggerates the scale dimension of the space. And the concise ceiling emphasizes the order of the space. The perfect integration of modern fashion perception and oriental elements is deeply rooted in the whole space and contains magnificent and profound oriental artistic perception. This is in accordance with the current aesthetic interests.

The design draws inspirations from oriental traditional elements and courageously makes some "destructions" and "negations", thus to create some brand new design concepts. The designer is good at finding out design elements having abstract expressions from the traditional symbols, thus to touch the most sentimental zone in the heart. The designer ingeniously puts Chinese style screen in front of the French window as the shelter. Sunshine shines through the gaps to make the screen appear glistening and arouse the vigor of the space. The sitting area is confined with round objects which not only have intensive outlook, but also can make people feel the harmonious feeling of the mutual integration of space environment. Behind the expressive forms of culture and art, the designer would rather express the power of art which virtually strengthens the mysterious tension of the space.

Neo-Classical Reflection·Top Leisure Business Clubs

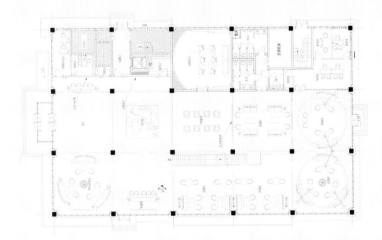
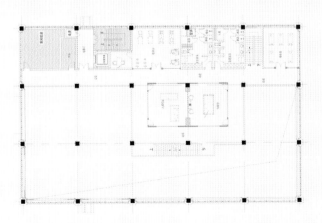

Oriental Ginza Club 东方银座会馆

设计单位：深圳市盘石室内设计有限公司	Design Unit: Shenzhen Panshi Interior Design Co., Ltd.
设 计 师：吴文粒	Designer: Wu Wenli
项目地点：辽宁省盘锦市	Project Location: Panjin in Liaoning Province
项目面积：3000 m²	Project Area: 3000 m²
主要材料：意大利木纹大理石、皮革、壁纸、艺术玻璃	Major Materials: Italian Wood Grain Marble, Leather, Wallpaper, Artistic Glass

在本案中，室内设计的基调以简洁、干净，为空间营造温暖气息为主。设计中所用的材质简单大方，大面积的使用钢化玻璃作为楼梯扶手的隔断，让空间显得通透、灵动。沙盘后方的屏风演绎着中国古典元素的美，隔断上一块块的红色仿佛漂浮在空中，形成装饰空间色彩的一个个点，生动而跳跃。

业主希望可以打造出有别于一般销售中心的商业氛围，让人仿佛置身于顶级Coffee Shop般的幽雅环境。因此，设计师通过改变过去"以使用功能及装饰为主"的操作模式，在空间中置入多种装置艺术，试图营造出一种新鲜、有活力的氛围。另一方面，也借由这些装置艺术，重新串连起人、活动与场所之间的连结，进而更主动地去拉近人与人之间的距离，触发人对空间不同的感受。

For this case, the tone of the interior design is conciseness and clarity to create warm atmosphere for the space. The materials selected in this case are mostly simple and decent. Large area of tempered glass was used as partition for staircase railings to make the space appear transparent and active. The screen behind the sand table displays the nice parts of classical Chinese elements. Blocks of red on the partition are like floating in the air, forming one after another spots decorating the color of the space, very active.

The property owner hopes to have some business atmosphere different from ordinary

sales center to make people feel like being in the elegant environment like top Coffee Shop. Thus, the designer turns from the former approach modes of centering on application functions and decorations and sets various decorative artistic objects in the space to create some fresh and active atmosphere. On the other hand, these decorative artistic objects reconnect people, activities and sites, and shorten the distance between people further, arousing different experiences of people towards the space.

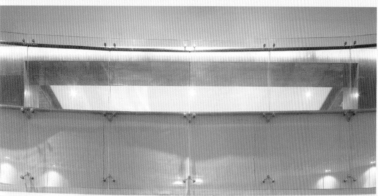

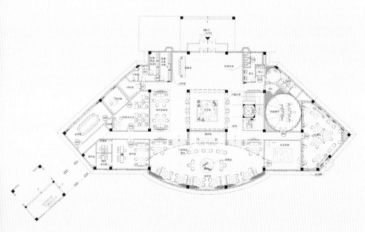
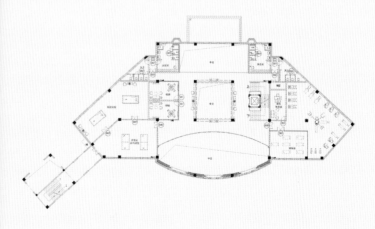

Sunshine Gold City Villa Club

阳光金城别墅会所

设计单位：PINKI 品伊创意机构＆美国 IARI 刘卫军设计师事务所	Design Unit: PINKI Innovation Institution & IARI Liu Weijun Design Firm
开发商：西安国中星城置业有限公司	Developer: Xi'an Guozhong Xingcheng Properties Co., Ltd.
设计师：刘卫军	Designer: Liu Weijun
项目地点：陕西省西安市	Project Location: Xi'an in Shanxi Province
项目面积：600 m²	Project Area: 600 m²
主要材料：浮雕面实木地板、壁纸、文化石、红洞石、古铜斑木饰面、米黄大理石等	Major Materials: Relief Solid Wood Floor, Wallpaper, Cultural Stone, Red Travertine, Bronze Wood Veneer, Cream-Colored Marble

Neo-Classical Reflection·Top Leisure Business Clubs

本会所规划为独栋别墅,是豪宅项目的重要组成部分之一,会所将成为提供给客户享受豪宅生活的重要标志。设计师将其视为勾勒西安上流阶层生活面貌的显要之居,既是西安一颗最耀眼的珍珠,更是现代都会内一处文化汇聚、视野开阔的绿洲!

设计师依据项目的高端性及客户群体的品味与身份的象征意义,考虑到生活起居和对外的宴客需求,将平面配置分为3层:一层为开放性的私人会所;二层为私密性的起居空间;三层为私密性休闲及写作空间。

设计师所设计的私人会所接近于一个招待会所或艺术展厅的概念,进入会所之后,经过配有换鞋间功能的门厅,首先来到以洽谈为主的副厅及客厅,三面共约29m长的落地窗的环景视野,铺陈副厅及客厅的宽阔尺度,副厅及客厅之间的景观水池起着功能分区及过渡的作用,利用手扶梯位置设计了文化石铺设的主题墙,让副厅及客厅的洽谈范围更具亲和力。通过主题墙的门洞就是可容纳6~8人的早餐厅及可容纳12人的餐厅,餐厅旁边配有22 m^2 的大厨房,体现了业主对烹饪的高度要求及对招待来宾的热情细致。通过餐厅边上设置了人性化的残疾人通道可直达水吧区,可在此品酒享用雪茄等。整体而言,一层私人会所区域包括一道围绕极佳采光的中心庭院廊道、水吧区、早餐厅、主餐厅、家庭影院、午茶区(木平台)等,从传统大而无挡的制式排场概念中跳出来。全面性的生活配套,配备齐全的招待会所及人文气息浓郁的交流氛围,给人们

带来静雅、闲适的生活情趣。

进入私密起居空间，可通过与中心庭院相连的楼梯和餐厅旁边的升降电梯两个选择。居住空间注入了更多的中西混搭风，两间次卧都配有独立的衣帽间及卫生间，透过两间次卧共有的大露台可观赏园区景色。在公共区域分别设有主家庭厅及副家庭厅，主家庭厅装饰以强调舒适温馨为主，是家庭成员主要的休息及交流中心，次家庭厅主要是供主人休息及接待密友所用。通过家庭厅可进入功能配套齐全的主卧，包括有：男女主人各自独立的衣帽间、洗手台及化妆台，还有配有直径约1.6 m浴缸干湿区分的洗浴空间。主卧同时还拥有内、外露台。内露台是业主主要的休闲空间，可通过此露台与在一层的中心庭院的朋友或家人进行交流，外露台可观赏园区景色。通过主卧小门旁的楼梯可方便地到达业主三层私密性休闲及写作空间，主要延续了一层追崇自然的山居设计风格，配有洗手间、书房、瑜伽室、内露台、外露台。

设计师认为中国人的审美思维是一种对应关系，因为中国人长期从自然中观察体验出来的一种相对事物的观念与现象。而这种观念作用在创造人为空间时，采取与大自然相呼应、相对话的关系，而不是一种征服的方式。因此在空间序列安排、外界的环境呼应，以及细节布置的陈设上，都采用相互呼应的手法以现代烘托中式典雅、以简约衬出细腻繁美，以此来表达现代中国人文化意识的特点。

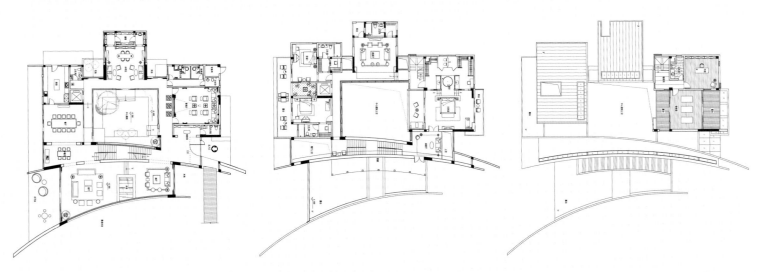

This club is a single villa and is an important constituent part of the mansion project. This club would become the key symbol for customers to enjoy mansion life. The designer considered it to be an important place to display the living conditions of elite class in Xi'an. It not only is a glaring pearl in Xi'an, but also is an oasis with broad views which can also integrate cultures.

Depending on the high-end characteristics of the project and the symbolic significance of the taste and status of the customers, taking into considerations the customers' daily life and their requirements, the designer divides the floor configuration into three layers: the first floor is an open private club, the second floor is private living space and the third floor is set for private leisure and study space.

The private club by the designer is something like a reception club or an artistic exhibition hall. Upon entering the club, passing the hallway where one can change

shoes, one would first come to the auxiliary hall and living hall designed for negotiation purpose. The three sides 29m long French window has somewhat panoramic view. Both these halls have broad scale. The landscape pool between these two halls can be functioned as division zone and transition zone. The theme wall paved with cultural stone is designed along the staircase, which make the negotiation atmosphere in these halls have much more affinity. After you walk across the door opening in the theme wall, you can come to the breakfast dining hall which has a capacity for 6~8 people and a dining hall for 12 people. 22 m^2 kitchen beside the dining hall represents the high demands of the owner and the warm reception towards the guests. The disabled access along the kitchen can lead to the water bar area where you can enjoy wine or smoke cigar here. The first floor's private club area includes central yard gallery with fine lighting, water bar area, breakfast dining hall, main dining hall, home theatre and afternoon tea area, etc. The complete living accessories, reception club and the interaction atmosphere with intense human atmosphere create some elegant and leisurely living interests for people.

If you want to come into the private living space, you can choose to use the staircase which connects with the central courtyard or you can use the elevator beside the dining hall. The living space is instilled with some more east and west mix and match style. The two guest bedrooms are equipped with independent cloakrooms and washing rooms. Through their shared terrace, you can enjoy

the garden landscape. There and main living hall and secondary living hall in the public area. The decoration of the former one emphasizes on comfort and warmness and is the main lounge area and communication area for the family members. And the latter one is mainly for the owner to rest and receive close friends. Through the family hall, you can come to the master bedroom with complete functions and accessories, which mainly includes the host and hostess' respective independent cloakroom, washbasin and dresser. Other than that, there is the bathroom which separates the wet area from the dry area and has a bathtub which is about 1.6m in diameter. The master bedroom has an interior terrace and an exterior bedroom. The former one is the master's main leisure space and he can communicate with friends or family members in the first floor's central yard via this terrace. One can enjoy the garden views on the exterior terrace. Through the staircase by the side door of the master bedroom, one can easily get to the private leisure space and study space on the third floor.

It mainly continues the mountain residence design style of the natural first floor. The third floor is equipped with bathroom, study, Yoga room, interior terrace and exterior terrace.

The designer believes that the aesthetic thinking of Chinese people is some kind of corresponding relationship. This is some attitude and phenomenon that Chinese people accomplished from the long history towards things in the world. While creating human space, this kind of attitudes applies the relationship in correspondence and dialogue with nature, rather than a conquering way. Thus as for space order arrangement, echoing the exterior environment and the layout of details, the designer makes use of mutual correspondence and sets off traditional Chinese elegance with modernity, delicacy with conciseness, thus to represent the characteristics of modern Chinese cultural conscience.

Huizhou Golf Club

惠州高尔夫会所

设计单位：KSL 设计事务所
参与设计：林冠成、温旭武、马诲泽
项目地点：广东省惠州市
项目面积：10000 m²
主要材料：樟木、秀石、皇家木纹大理石、皮革、斑马木饰面、灰木纹大理石、黑钢、特殊玻璃、石材陶瓷锦砖

Design Unit: KSL Design(HK) LTD.
Associate Designers: Andy Lam, Wen Xuwu, Ma Huize
Project Location: Huizhou in Guangdong Province
Project Area: 10000 m²
Major Materials: Camphorwood, Imperial Wood Grain Marble, Leather, Zebra Wood Veneer, Grey Wood Grain Marble, Black Steel, Special Glass, Stone Mosaic

惠州高尔夫会所坐落于广东省惠州市惠东县大岭镇，地处高尔夫球场及湖边，风景秀丽，地理位置极佳。KSL设计事务所的设计力求让这个空间更加静谧而有深度，对空间的功能分区进行了合理的布置，让客人无论休息或是用餐时，均能将高尔夫球场的美景尽收眼底。

项目的室内设计与建筑外观的中式风格相呼应，以新中式的折中手法，将具有国际感的现代家具及艺术装饰完美组合在一起，达到空间视觉上的统一。设计师注重质感的表达，希望在东方情韵浓厚的室内环境之中，也能够体现出西方人本主义的舒适与优雅，更是让客人在休闲舒适之外，能感受到内在的奢华与品位。质朴的材料和恢弘的空间，简约的装饰与精致的氛围，极致的对比给人以融汇中西、时空错落的非凡享受，更在山水相依的设计中，升华了人生的境界。

This project is located in Daling Town, Huidong County, Huizhou in Guangdong Province. This location along the golf course and lakeside is very picturesque and has wonderful geologic location. The design by KSL Design Firm tries to make this space much more serene and profound and makes proper arrangement towards the functional division of the space. Thus when the guest is having a rest or having meals, they can always enjoy the sceneries of the golf course.

The interior design of the project echoes the Chinese style of the architectural outlook. With new Chinese style approach, the designer perfectly integrates modern furniture with international feel and the artistic decorations, thus achieving the visual unity of the space. The designer emphasizes on the expression of the texture and hopes that the interior environment with intensive oriental charms can also display the comfort and elegance of western style. What is more, the customers can feel the inner luxury and taste from this leisure and comfort. The primitive materials and magnificent materials, concise decorations and delicate atmosphere and extreme contrast give people incomparable enjoyment integrating east and west, which is like in totally another world. Other than that, the design of mountain and water uplifts the realm of human world.

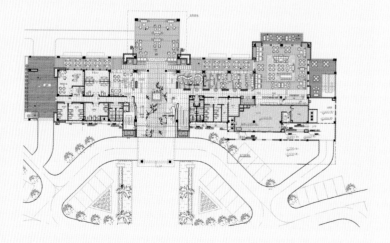

榕都晶华
RongduJinghua

设计单位：维思张开旺设计事务所
设 计 师：张开旺
参与设计：林文
项目地点：福建省福州市
项目面积：3000 m²
主要材料：木饰面、大理石、烤漆玻璃

Design Unit: Weisi Zhang Kaiwang Design Firm
Designer: Zhang Kaiwang
Associate Designer: Lin Wen
Project Location: Fuzhou in Fujian Province
Project Area: 3000 m²
Major Materials: Wood Surface, Marble, Baking Varnish Glass

走进安泰河畔的榕都晶华，目之所及的视觉体系恰似一曲慢板的前奏，迎接着人们的到来。这里带给人们一种很奇妙的感觉，仿佛从现实的喧闹中走来，在这个暖色调的空间里可以渐渐隐去那份浮躁。眼前错落分布的东西方元素与空间存在着微妙的关系，似乎已超出了单纯的存在意义。而在无中心的构图观念影响下，让其由里而外晕出淡雅、不张狂的光影，释放着开朗的气质。

将大小不一的人偶雕像摆放在层次感十足的大理石台面上，像似一场表演的开场白，带给人无限的期许。其背后的休憩区以混搭的形式衍生出别样的视觉质感，而一旁的墙面则是曲线的造型，强化了整体的韵味。沿梯而上，空间走道的墙面上也适时地添加了后现代的雕塑品，而一些中式的摆件也穿插其中，差异化的解构颇有戏剧性。人们可以对这些物件有各种各样的猜想，存在即是合理。

榕都晶华既有独立的包厢，也有位于公共区域的坐席。前者用细腻的空间表达给予人们消费的归属感，后者采用通透的材质做隔断，呈现开放式的布局效果。榕都晶华向我们展示了小资的情调和时尚的氛围，同时又满足了消费者对场所的幻想与期待。

在这个功能主义与表现主义并重的空间里，我们很难用"最"字去形容某个场景，因为你可能刚为眼前的一个设计而赞叹，转身在某个拐角或是眼光不经意间地掠过，遭遇的又是另一种经典。渐渐地，榕都晶华里的每一个个体都有着属于自己的梦想，而作为食客的我们，也似乎融入了空间，举手投足、谈吐皱眉间都像是电影的片段被如实地记录。

Neo-Classical Reflection·Top Leisure Business Clubs

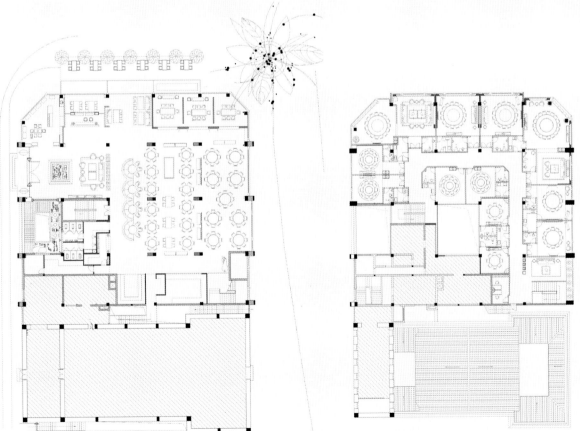

Once you enter this club by the side of Antai river, what comes into the eyes is like some preludes, welcoming people. This place gives people some very strange feeling, like coming from the noisiness of reality. In this space with warm color, people can gradually get away from the fickleness. The oriental and western elements scattered at random have some subtle relationship with the space, going beyond the pure existence significance. Influenced by centreless composition attitude, the interior and exterior space displays some elegant and reserved light shadow, with some cheerful temperament.

On the marble surface with rich layers, there are big and small figure statues, which are like opening remarks for a performance, creating limitless expectations for people. The lounge behind produces some peculiar visual texture in some mix and match way. And the wall beside strengthens the whole charms with curved formats. When you walk upwards along the stairs, there are post-modern sculpture objects on the wall along the corridor. There are also some Chinese objects among them. This construction with differences is quite dramatic. People can have various imaginations towards these objects. What exists is reasonable.

This club not only has independent boxes, but also has sitting area in the public area. The former ones give people sense of belonging through meticulous space expression. The latter one uses transparent materials as the partition, displaying the effect of open layout. This club presents to people bourgeoisie charms and fashionable atmosphere, while at the same time meeting people's imaginations and expectations towards a club.

In this space emphasizing both on function and on expression, it is hard to describe some setting with the word "most". It is because that for this moment you would get amazed by one design in front of your eyes, around the corner, you would come into some other classical design. Gradually, everybody here has his exclusive dreams. As people enjoying meals here, we are like to get integrated into the space. All movements are recorded like segments of a movie.

Haitong No. 1 Mansion
海通一号

设计单位：福州国广装饰设计工程有限公司
设 计 师：丁培瑞
项目地点：福建省福州市鼓楼区
项目面积：900 m²

Design Unit: Fujian Guoguang Architectural Decoration Design and Engineering Co., Ltd.
Designer: Ding Peirui
Project Location: Gulou District, Fuzhou in Fujian Province
Project Area: 900 m²

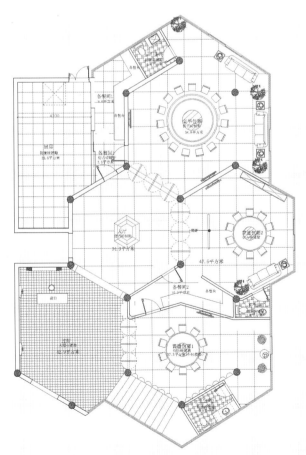

本案运用精彩的空间视觉语言，希望人们在一种从未体验过的奢华环境中犒劳自己的味蕾。在大堂区域，瓷砖地面的纹理在曲折变化中，柔化了材质的硬朗特质，并配合着一侧传统中式韵味的家具与装饰品，增添了亲和力。而顶棚则由层层叠叠立体的金色造型来衬托，如花朵、似浮云。

包厢中，用鲜艳明快的色彩来降低中式风格的沉稳，使其焕发崭新的面貌。朴素的灰砖石地面在现代水晶灯的衬托下，衍生出别样的气质。传统优雅的仕女图与各式窗格相得益彰，展现出一种多层次的美感，让空间视觉体验更加丰富。而那些传统线条的桌椅和精致的布饰相互呼应，削弱了中式家具的沉重感，反而令整个空间有一种清新的气场。

除了装饰上体现的奢华精致，"大空间"的概念也是这里设计的重点。用餐区、等待区、休闲区等功能空间区分明确。而光影的微妙变化所创造的虚实互补，虽然并不张扬，却能潜移默化地影响着各个区域的气质，带给客人温和、内敛、充满人情味的消费氛围。希望通过对传统文化的深刻理解，将色彩与造型进行有机的结合，再以崭新的角度诠释传统文化。让整个空间如同一壶茗茶，洋溢着一股清新的气息，令人流连忘返。

This case applies excellent space visual language to please visitors' taste bud in this luxurious environment that they have never experienced before. In the lobby, the texture of the ceramic tile softens the tough characteristics of the materials and adds some more affinity accompanied by furniture with traditional Chinese charms on one side. The ceiling is set off by tiers upon tiers of three-dimensional gold modeling, like flowers or flowing clouds.

Inside the box, the designer uses brisk and bright colors to make this Chinese style appear not that sedate and displays its brand-new appearance. The plain grey masonry ground is set off by the modern crystal light creating some peculiar temperament. The traditional elegant portrait of ladies and various kinds of windowpanes bring out the best in each other, presenting some multi-layered beauty and enriching the visual experiences of the space. And those tables and chairs with traditional lines and the exquisite cloth echo each other which weakens the heavy feel of traditional Chinese furniture, but bring some fresh atmosphere to the whole space.

Apart from the luxury and delicacy in the decoration, the concept of "big space" is also the focus of the space. There are distinct separation among dining area, waiting area and leisure space. The subtle change of light and shadow creates some complementation between what is real and what is unreal. Although it is displayed in a wild way, it can quietly influence the temperament of every zone, thus bringing to guests some mild, moderate and human consumption atmosphere. The designer hopes to combine colors with models organically through profound understanding towards traditional culture. And then he can interpret traditional culture from a new angle. The whole space is like a pot of tea with fresh smell. People would enjoy themselves so much as to forget to leave.

水煮工夫茶道会所

设计单位：昊辉空间设计事务所	Design Unit: Haohui Space Firm
设 计 师：曾昊	Designer: Zeng Hao
项目地点：福建省宁德市	Project Location: Ningde in Fujian Province
项目面积：1600 m²	Project Area: 1600 m²
主要材料：杉木、LOLA 瓷砖、天然花岗岩、陶土砖、原生木蜡油	Major Materials: China Fir, LOLA Ceramic Tiles, Natural Granite, Pot Clay, Wax Oil
摄 影 师：周跃东	Photographer: Zhou Yuedong

本案位于福建省宁德市东侨区南岸景观公园内，原址为宁德市首届茶博会会场。

为满足功能要求，全场分成茶艺休闲区和就餐接待区。大量中式回廊的运用，有机地融合了两大功能区，使之在相对独立的基础上依旧能保持气场上的完整统一，并且为来宾在内场走动线路上提供客观引导作用。借用中国传统手法中的水榭楼阁、窗花月洞、斗拱藻井、流水连廊等各式形态，引发老子思想无中生有之道。设计上有意识地虚空大堂部分的顶面造型及地面陈设，使整体空间所呈现的气质，成功规避了骄奢媚俗之态，代之以空灵、儒雅、别致的大家之气。

迎合人居环境低碳环保的设计理念，本案选用了经济而环保的杉木原材、天然花岗岩、陶土砖、原生木蜡油等材料，贯穿全场，让空间纯朴自然，浑然天成而不失端庄优雅。色彩上以原材料本色为主导色，配以传统朱红廊柱、彩绘画梁，成功再现了浓浓的东方风韵。

This project is located inside Southbank Landscape Garden of Dongqiao District, Ningde in Fujian Province. This place used to be the meeting place for the first tea exhibition of Ningde City.

According to the requirements of functions, the whole site is divided into tea leisure area and dieting reception area. The application of many Chinese style winding corridors organically integrates two functional areas. Thus the space can maintain environmental unity on comparatively independent foundations. And this arrangement can provide the guests with leading ways. Acquiring various forms in traditional Chinese approaches, such as waterside pavilions, bracket system, corridors, etc., all these arouse the philosophy of Lao Zi from people's heart. As for design, the designer consciously makes partial ceiling and ground decoration void. Thus the temperament of the whole space successfully avoids the situation of luxury and vulgar which is replaced by airy, elegant and peculiar atmosphere.

In order to meet the design concept of human setting environment and low-carbon, in the selection of materials, this project chose ecological and environmental China fir, natural granite, pot clay, wax oil, etc. throughout the space to make the space pure and natural, like nature itself, but dignified and elegant. In the color selection, it has the raw material color as the tone color, accompanied with traditional scarlet posts and painted beams. Thus intensive oriental atmosphere is represented.

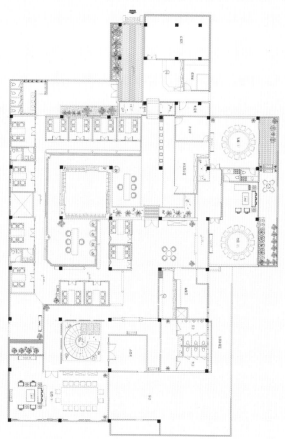

Hakkas Impression 印象客家

设 计 师：陈杰	Designer: Chen Jie
项目地点：福建省福州市	Project Location: Fuzhou in Fujian Pronince
项目面积：800 m²	Project Area: 800 m²
摄 影 师：周跃东	Photographer: Zhou Yuedong
撰　　文：江雍箫	Composer: Jiang Yongxiao

任何一种文化、一种理念，都要通过一个载体来培养，发扬光大。印象客家便是这样一个地方，它在设计师的精心规划之下于有形无形之间塑造出许多耐人寻味的情境。于是，我们在此用餐或品茶，体验到的不仅是味蕾的高级享受，更是一个触动心灵的过程。

印象客家位于 A-ONE 运动公园内，隐于深处的地理位置给这个餐饮空间多了几分低调与内敛。"追根溯源，四海为家"的文化理念也在潜移默化中得到些许诠释。印象客家的门面上方用斑驳的铁皮做装饰，粗犷的纹理显得厚实而有力量感。下方的圆窗位置，摆放着石磨与擂茶饼，墙面上的地图指示出客家族群在国内的分布情况，这些与客家文化一脉相承的物件在这古朴的空间中悠悠不尽。

尚未进入空间内部，外面的庭院景观已然吸引了我们的目光，曲径有秩的布局丰富了视觉的层次，得益于此，设计师在这个环境中设置了若干包厢。包厢置于自然的怀抱之中，食客便拥有了广阔的视野。同时，玻璃墙面使得窗外郁郁葱葱的景致成为一道天然的背景。渐渐地，这里的一草一木、一砖一瓦，不管是有生命的还是没生命的，都找到了与空间沟通共融的方式。

进入二楼的区域，玄关墙上的石壁彰显着客家的文化特性。石壁两侧分别是饮茶区与就餐区，中间的隔断便是从一楼水景处延伸上来的部分。两个功能区域之间可以相互借景，而不是硬性的区分。

印象客家虽是一个质朴的空间，但这种质朴并非奢华的对立面，而是一种平凡的表象，骨子里却充满了丰富的情愫。当阳光顺着周遭的树木摇曳而下，而后暖暖地映在包厢中，不用过多的修饰，如同一张讲述闲适情怀的电影海报。而夜幕降临时，在暖色的灯光映射下，空间便富有了诗意，充盈着迷人的气息。

Any culture, any concept, all need some carrier to be fostered, and be carried forward. This club is such kind of place. Under elaborate planning, the designer creates some enchanting circumstance. While we enjoy some dinner or taste some tea here, what we experience is not only the senior enjoyment for the taste bud, but also a process to touch the heart.

It is located inside A-ONE sports park. This geological location in the depth creates some low-key style and powerful connotations. The cultural concept of "tracing the root and origin, and home today and gone tomorrow" was interpreted in some subtle unconscious way. The space above the door surface was decorated with mottled iron sheet. The straightforward texture appears solid and powerful. In the location of the under round window, there is a tone mill. The map on the wall indicates the distribution of Hakkas ethnic groups in China. These objects representing Hakkas culture appear much more profound in this space of primitive originality.

Before you enter the interior space, the eyesight is already attracted by the exterior courtyard landscape. Meandering and orderly layout enrich the visual layers. Based on this, the designer sets quite a number of boxes inside the environment. With the boxes set in the natural embrace, the guests can have broad views. Gradually, for the tree and plant, every brick and tile, be it animate or inanimate, all find a way to communicate and integrate with the space.

For the area on the second floor, the stone on the wall of the vestibule displays the cultural characteristics of the Hakkas. On both sides of the stone were tea area and dieting area. The central partition is the part continuing from the waterscape. Two functional areas can borrow sceneries from each other, rather

than hard distinction. Inside the dieting boxes, Chinese

Although Hakkas is a primitive space, this primitiveness is not the opposite of originality. It has ordinary outlook, but inside it is abundant with rich sincerity. When sunshine swayed down along the surrounding trees, and warmly reflected in the box rooms, it is like a movie poster with leisurely sensations which does not need many embellishments. While night falls, reflected by warm lights, the space becomes much more poetic, full of fascinating atmosphere.

Sanming Club 三明会所

设计单位：福建品川装饰设计工程有限公司
设 计 师：周华美
项目地点：福建省三明市
项目面积：1125 m²
主要材料：皮革、壁纸、大理石、铁刀木、银箔

Design Unit: Fujian Pure Charm Decoration and Design Co., Ltd.
Designer: Zhou Huamei
Project Location: Sanming in Fujian Province
Project Arear: 1125 m²
Major Materials: Leather, Wallpaper, Marble, Indian Rose Chestnut, Silver Platinum Alloy

在这个融合了中式古典、现代简约与新古典风格的空间里，唯美的古典元素被融入到现代的风格中，稳重、精致的色调及古典图案作为主轴线贯穿了整个空间。设计师用纯净、淡雅、明快的色调作背景，衬托高品位的家具、灯具及艺术品陈设，结合突出的立体感与节奏感，烘托出会所高雅的文化氛围，这也成为贯穿会所各区域设计的基本准则。会所整体空间的装饰装修风格包容古今，设计师擅于在变化中寻求结合点，并努力贯彻到内部空间中，力求在追求整体统一的风格中体现多元化、多视野的文化内涵。

基于这些高度统一的设计理念，使得会所中的整体装饰呈现出豪华与高雅并重的装饰效果。此外，设计师在掌握全局装饰风格基调的基础上，又兼顾了开放、大众化，同时又保持私密、个性的环境。使得人们无论是进入大堂，还是步入包厢，都能感知到其观赏性与艺术性并重的画面。

会所的整体空间采用大面积的大理石、壁纸及木质屏风隔栏，使得高贵优雅的气质呼之欲出，同时也让人欣赏到线与块面的流畅性。各种装饰元素的充分应用，让空间充满活力，使得大自然的气息也灵活穿梭其间。

In this space integrating classical Chinese, modern concise and neo-classical style, aesthetic classical elements were integrated into modern style. As the main axis, modest and exquisite color tones and classical diagrams run through the whole space. With pure, elegant and brisk color tone as the background, the design sets off high-taste furniture, lights and artistic layouts. Combined with highlighted sense of three-dimension and rhythms, the elegant cultural atmosphere of this club was displayed. And this becomes the basic principle for various spaces' designs of this club. The decoration of the whole club space incorporates ancient and modern style. The designer is good at searching for bonding points in these variations, and implements them into the interior space, thus to represent cultural connotations of diversification and multiple visions in attaining integral style.

Based on these unified design conceptions, the whole decoration of this club displays some decorative effects emphasizing on both luxury and elegance. Other than that, based on mastering well the whole decorative style, the designer pays attention to open and public style, while maintaining some private and personal environment. This makes it possible that people can perceive a picture of focusing on both outlook and artistic sphere while entering the lobby or the box rooms.

The whole space of the club applies large area of marble, wallpaper and wood screen partition, thus the noble and elegant temperament is vividly portrayed. At the same time, people can perceive the fluency of lines and surfaces. The sufficient application of various decorative elements makes the whole space appear more active and the atmosphere of nature is present in the space.

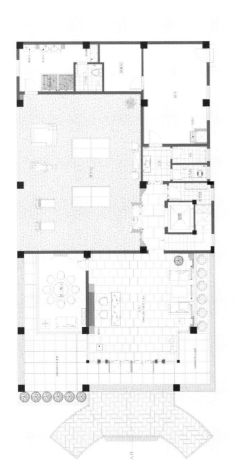

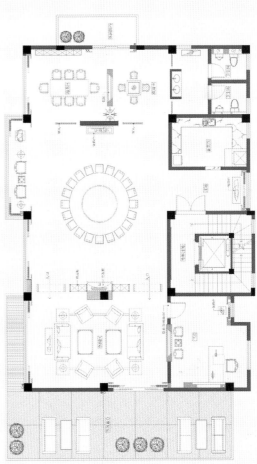

茗仕汇·茶会所
Ming Shi Hui · Tea Club

设 计 师：陈杰
项目地点：福建省福州市
项目面积：400m²
摄 影 师：周跃东

Designer: Chen Jie
Project Location: Fuzhou in Fujian Province
Project Area: 400m²
Photographer: Zhou Yuedong

一具石像静伫在小径尽头，仿佛已等了许久，让此处像是心的故乡之所在。青砖墙上生长的绿植生机勃勃，也有耐人寻味的所指。顶棚上奇异的灯如同朦胧的月亮，只是换了修长的身姿，将来者的思绪拉得绵长而愈发朦胧。

古意盎然游离于经纬之间，一方清雅使得静安于市，闲坐在此，约三两好友品茗长谈，何止是难得，亦是惬意。一盏清茶，几句调侃，呼朋唤友，好不自在逍遥。设计师便是为了如此这般的缘由，用其巧手妙思，圆了我们一个禅意十足、意味颇丰的品茗梦。

推开此扇方圆和谐的门扉，将喧嚣与浮躁都留给了两旁翠竹，前往这淡然的所在，去体会那久违的轻松。走过青石铺就的路面，苍劲的书法在浅薄的卷面上翻飞，光透过以此为屏风的隔断，令这些墨迹泛出久远的意蕴，充满摄人心魄的美。白色的鹅卵石散落在路边，交错平行的立面像

把折扇慢慢推开，将空间的意蕴荡漾开去。而人眼尽是简而有味的明式隔断和家具，令古意弥漫在空间每个角落。拱形的青瓦垒成的隔断形成独特的美感，光透过瓦片之间的空隙照射进来，明暗之间好像藏着又一个天地。配合着墙面的纹理，枯木之上放置的一小盆植物，蕴含了那份枯木亦逢春的哲思。简洁的线条，给予空间纯粹的力度与美感，可谓实用性与美学的完美结合。而那满墙的各式雕花，令人为之动心，想必是花了不少心思收集而成的，怎能不令人为之折服。那一把古琴弹奏的又是哪一曲意味深长的古调呢？闲坐在此，喝一杯清茶，听一首古曲，可以忘却那流水般逝去的时光，这也是一件使人倍感满足的事情。

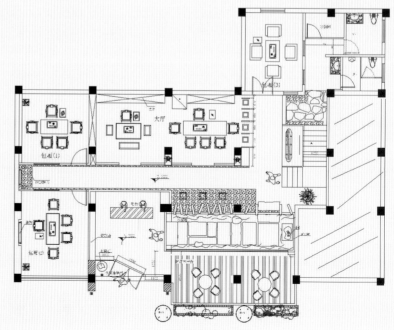

A stone sculpture stands quietly at the end of the alley. It is like it has been there waiting for quite a long time, and it is like here is the hometown for heart. There are green plants full of vigor growing on the black brick wall which is quite enchanting. The bizarre lights hanging from the ceiling are like hazy moon. It is only that it has slender outlook which elongates and dims the visitors' thoughts.

All through the space you can find antique artistic conceptions. This space of elegance makes people enjoy the serenity here. During the leisure time, you can sit here with several friends, enjoying tea and having a long talk. This is not only some rare occasion, but also that you can have some satisfaction. You can enjoy some green tea, laugh and talk with several friends. What an easy and happy time. Just out of this reason, the designer ingeniously fulfills our dreams full of Zen style and meaningful connotations.

After you open this door with harmony, and go into this quiet location experiencing this long-lost relaxation while leaving hustle and bustle to the verdant bamboos on both sides. After you walk across the road paved with bluestone. Hard and strong calligraphy is rolling on this paper. Light penetrates through the partition with this as the screen. These ink marks have some lasting connotations, with fascinating beauty. White cobblestone lies on the roadside. Staggered and paralleled façade is like a folding fan. When you open it, the connotations of the space would spring outside. You can find simple and profound partition and furniture of Ming Dynasty style. And the antique sensation is pervasive throughout the space. The partition of arched grey tile forms some peculiar sense of beauty. Light penetrates inside through the interval between the tiles. It is like there is another world between light and darkness. Accompanied with the texture of the wall, there is a small pot of plant on the top of the dead wood, containing the implication of dead wood sprouting in spring. The concise lines give the space some pure strength and aesthetic feeling, combing perfectly practicality and aesthetics. All kinds of carvings on the wall arouse people's enthusiasm and interest. It must have cost the designer quite a lot of time and thinking to collect all these. You can enjoy a cup of green tea or listen to some ancient piece of music. You would forget about that time flowing away like flowing water. What a pleasing pastime this is!

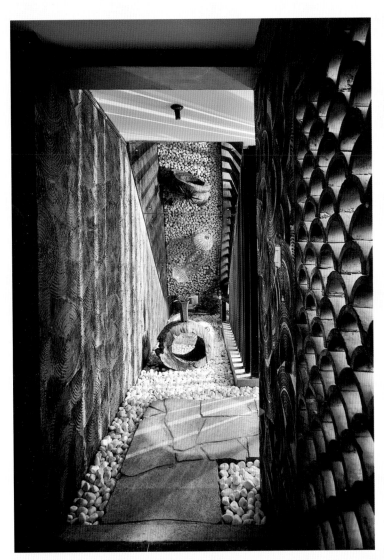

First Talk Dark Red Enameled Club

第一谈紫砂会所

设计单位：道和设计机构	Design Unit: Daohe Design Firm
项目地点：福建省福州市	Project Location: Fuzhou in Fujian Province
项目面积：116 m²	Project Area: 116 m²
主要材料：黑钛、黑镜、壁纸、木质花格、木纹砖、仿古砖、毛石	Major Materials: Black Titanium Steel, Black Mirror, Wallpaper, Wood Lattice, Wood Grain Brick, Rustic Tile, Quarry Stone
摄影师：李玲玉	Photographer: Li Lingyu

本案位于滨西66号的内部，处于一个安静恬逸的环境中。入口处，细致的不锈钢字落于粗糙的毛石碎拼墙上，两种不同的材料质感对比强烈，显示出质朴和低调。旁边的木质格栅引领着人们的目光进入室内，在灯光的衬托下，一个个造型迥异的紫砂壶如同艺术品般被对待，这种如同博物馆的陈列方式，不仅能达到紫砂壶作为会所展示品的目的，又能将紫砂壶突出于整个环境之中。

设计以中式元素为主，茶室中古朴的明代圈椅隐隐流露出历史的沧桑感，简洁干净的背景墙搭配着水墨韵味的荷花，极能突出茶室素雅、安静之感。顶棚处悬挂着鸟笼状的灯具，不仅能融入室内环境，更是添加了些许趣味。最为突出的是，借鉴传统的中式雕花花格玻璃在室内的运用，让传统与现代相互交融，产生出不一样的视觉美感。

This project is located in the interior place of No. 66 Binxi which is a quiet and relaxing place. At the entrance, you can find meticulous Chinese characters on coarse rubber stone wall. A strong contrast is performed between these two different texture materials, displaying plainness and low-key style. The wood grating on the side leads people's eyesight into the interior space. Set off by the light, one after another dark red enameled pottery is like artistic object. This kind of library-like layout not only achieves the purpose of applying these potteries as club exhibits, but also highlights the status of these potteries in the whole environment.

The design is centered on Chinese elements. The primitive round-backed armchairs of Ming Dynasty style in the tea room reveal some vicissitudes of time. Succinct and clear background wall is accompanied with ink painting lotuses with lasting charms which highlight simple but elegant and quiet senses in the tea room. A birdcage like lighting accessory is hanging from the ceiling which not only integrates into the interior environment, but also adds some interests. What is more prominent is the application of traditional Chinese carving pattern lattice glass which combines tradition with modern style bringing our some peculiar visual aesthetic perception.

Dalian Yida Miaoling Club

大连亿达庙岭会所

设计单位：PINKI 品伊创意机构＆美国 IARI 刘卫军设计师事务所
设 计 师：刘卫军
开 发 商：大连软件园开发有限公司
项目地点：辽宁省大连市
项目面积：3956 m²

Design Unit: PINKI Innovation Institution & IARI Liu Weijun Design Firm
Designer: Liu Weijun
Developer: Dalian Software Park Development Co., Ltd.
Project Location: Dalian in Liaoning Province
Project Area: 3956m²

本项目主题风格定位为"法式风情风格",以艺术之名,布局上突出中轴线的对称,恢弘的气势,豪华舒适的空间。细节处理上运用了法式的廊柱、雕花、线条,制作工艺精细考究。

从门厅一进来的两侧分流,让空间功能区域能够更加合理地利用。顶棚采用圆拱形,与门的弧线进行着呼应。在以顶棚为中心的装饰纹样中散发着关于欧式传统生活的联想,地砖的呼应贯穿着整个空间。大厅中欧式的恢弘气势会使人停下脚步,静静欣赏。大理石的罗马柱体现了稳重、浑厚的质感,顶棚的欧式油画及天窗的通透修饰,以及地面采用的欧式地毯纹样,让人仿佛置身于罗马教堂,感受到欧洲传统文化带来的陶醉。华丽的水晶吊灯在中庭上散发着光芒,炫目精致的细节,把空间装饰得奢华、典雅。

在空间延续手法上,设计师具有匠心独到之处。使空间在不失高贵、典雅的贵族风格中,又进行着变化,使之远离乏味、单调,在重复中带着节奏和韵律。就像正在演奏着的一部华丽、高贵、典雅的欧式交响乐。又好像倘佯在卢浮宫中聆听着毕加索、梵高的心声,这里是美术馆、陶艺馆、图书馆、是怀有心灵追求的人们邂逅与相识的美妙之地。本项目为山地住宅,带来的是一种健康的生活方式。一般都是坐落在群山怀抱、水声潺潺的峡谷或是山涧。所选地域要求空气含氧量充足、水质好、植被丰茂、视野开阔。所有的建筑按照不同的地势、落差和自然景观存在,伴着自然而栖。采用传统的围墙式的庭院,使建筑坐落在自然的环境中,拥有整体、优美的天然环境。

The theme of this project is set as French style. In the name of art, the layout highlights the symmetry of axis, magnificent ambient, and luxurious and comfortable space. In detail, we employed French pillars, carvings and lines. We asked for very refined and meticulous craftsmanship.

The two-route arrangement entering the hallway makes the utilization of the space functional region more appropriate. Arched ceilings correspond with the arc of the door. The decorative pattern with ceiling in the center arouses some imaginations about European traditional life. The floor tiles echoes the pattern throughout the space. People would stop and feel the huge manner of European style manner. Marble Roman poles make the balcony a great place to overlook the scenery. The ceilings with European paintings, the skylights with transparent ornaments and European carpet pattern on the ground make people feel like being in a church of Rome, experiencing the intoxication of traditional European culture. Gorgeous crystal chandelier from court exudes its light, dazzling and exquisite detail, the elegant space being decorated in a luxurious way.

The designer is quite ingenious in his approach to space continuation. There are changes in the noble and elegant aristocratic style, far away from being boring and monotonous, possessing rhythm in repetition. It is like some gorgeous, noble and elegant European symphony. It is also like listening to the heart of Picasso and Van Gogh in Louvre. This is an art gallery, a pottery museum, a library, a perfect place for the encountering of people with spiritual pursuit. This is a residence located in the mountainous region which upholds healthy lifestyle. Generally speaking, this kind of residences is located among mountains, in the canyon with gurgling water or river valleys. The location demands high level of oxygen in the air, fine water quality, luxurious plants and broad views. All buildings exist according to different topography, different fall and natural landscape, coexisting with nature. The buildings are located in the natural environment with courtyard of traditional enclosures, owning whole and beautiful natural environment.

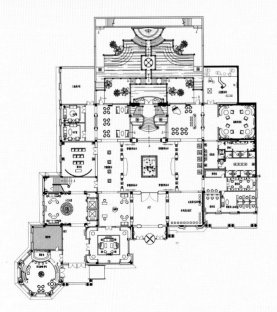
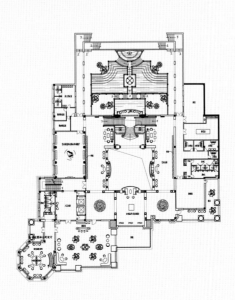
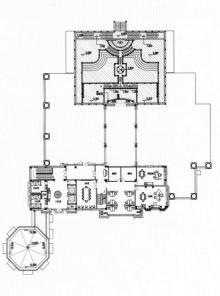

西溪·海会所
Xixi·Ocean Club

设计单位：元尚设计顾问（香港）有限公司
设 计 师：孙扬元
项目地点：浙江省杭州市
项目面积：约3000 m²
主要材料：大理石、陶瓷锦砖、木饰面、镜子
摄 影 师：支抗

Design Unit: SYY Design and Consultants (Hong Kong) Co., Ltd.
Designer: Sun Yangyuan
Project Location: Hangzhou in Zhejiang Province
Project Area: about 3000 m²
Major Materials: Marble, Mosaic, Wood Veneer, Mirror
Photographer: Zhi Kang

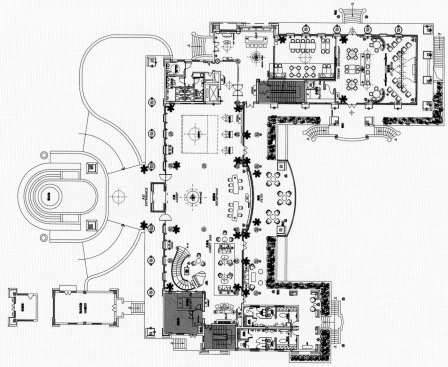

美国南部也经常被简称为"美南"、"迪克西"或直接称为"南部",构成了美国南部至东南部的一片广大地区。这个地区带有相当特殊的文化和历史背景,包括了早期欧洲殖民时期留下的痕迹,在装饰风格上有着明显的欧洲混合特点。

这种装饰风格崇尚人在舒适的环境里"优雅放松地享受生活"。挑空的白色木质顶棚,高大通透的窗户连接着户外,法式的百叶窗敞开着,使室外的景色可以与室内陈设相互呼应。装饰与实用功能兼有的吊扇,不仅能使空气流通,也给室内带来清新自然之感。装饰色彩及材料的选用源自于大自然的启示,明快的色调搭配天然的木砖、石材及皮革,使空间中洋溢着温馨气息。装饰手法在保留经典和传统的基础上强化舒适和宜居。家具、灯具在设计上注重保留传统的经典比例与材质。闪耀着光芒的法式水晶灯,英式、法式、意大利式的家具,舒适的面料,精美的艺术品,都自然地展现出空间的休闲或庄重。无论是开放式或是封闭式的空间设计,都充满了戏剧性的张力。

会所大堂的地面运用富有层次感的大理石拼花,配以精致的中心桌,结合色彩斑斓的花艺,在璀璨的水晶灯照耀下,显得富丽而高贵。整个空间米白色的色调,营造出明亮闲适的感觉。一层餐厅区用优雅的地面拼花,舒适的餐桌椅、长长的艺术造型镜面共同打造出空间的温馨感。阅览室温暖的木饰面搭配着舒适的家具、地毯,营造出安静的空间氛围。酒吧、钢琴、大大的餐桌及舒适的坐椅等,使会所真正的体现出"大家的大客厅"的亲切感。

Southern United States or American South were usually named Dixie or the South. It includes a grand area from the south of America all the way to the South-East of America. This region has some peculiar culture and historical background, such as trace of the earlier European colonization period. For the decoration style, it has apparent European blending characteristics.

This kind of decoration style makes it possible for people to "enjoy life in an elegant and relaxing way" in the comfortable environment. It has high white wood ceiling, and the transparent window connects with the exterior world. The French-style blind window is open so that the interior space can be made fresh and natural. The selection of decoration colors and materials were all enlightened by grand nature. The combination of brisk color tone and natural bricks, stone and leather make the space appear very warm. The decorative approach highlights comfort and livability while maintaining classical and traditional style. The design of furniture and lights accessories emphasize on the classical apportion and traditional materials. The shining French-style crystal lights, furniture of British, French or Italian style, comfortable materials and delicate artistic objects all display leisure and graveness of the space. Be it open or enclosed space design, all are full of dramatic tension.

The ground of the club's lobby makes use of rich-layered marble pattern, accompanied with delicate table and colorful floriculture. Illuminated by the dazzling crystal lights, all appear luxurious and aristocratic. The cream-colored tone in the whole space creates some bright and leisurely feeling. The warm wood veneer in the reading room, accompanied with comfortable furniture, carpet, creates a quiet space atmosphere. The wine bar, piano, grand dining table and chairs make this club display some intimate feeling of "a grand living room for everyone."

Huajingli Sales Club
华景里售楼会所

设计单位：广州共生形态工程设计有限公司／WWW.COCOPRO.CN
项目地点：广东省广州市
项目面积：730 m²
主要材料：大理石、玫瑰金色不锈钢、花梨木饰面、抛光砖、木地板等

Design Unit: Guangzhou C&C Design Co., Ltd./WWW.COCOPRO.CN
Project Location: Guangzhou in Guangdong Province
Project Area: 730 m²
Major Materials: Marble, Rose Gold Stainless Gold, Rosewood Veneer, Polishing Brick, Wood Floor

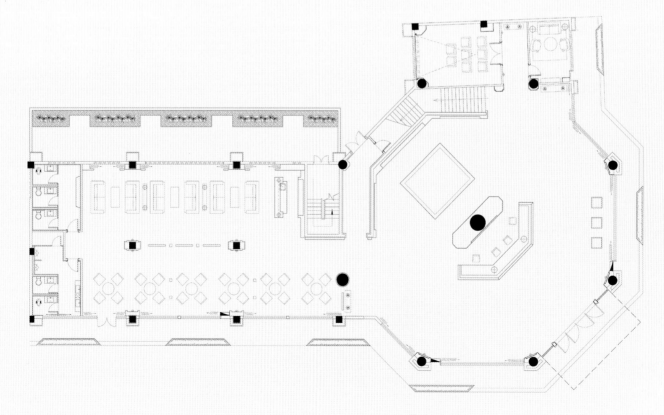

本案位于广州天河华景新城,设计风格为奢华的古典欧式风格,方案以突显售楼部的尊贵和大气,带给客人非一般的视觉体验为首要任务。八角楼形状的售楼部在功能分区上主要分为两个部分,前厅主要是接待大厅和沙盘展示区,后厅为洽谈休闲区。

洽谈区以镶着精美的铁艺图案的屏风作为隔断,让空间在展现精致的同时更不失通透。丰富多变的几何形地面石材拼花,精美柔和线条的视觉效果,配合典雅的家具与配饰,欧式的华贵与现代的舒适相融合、古典的柔美与现代的硬朗相碰撞,从而彰显出售楼部大气、优雅的高贵气质。

This project is located in Huajing New Town of Guangzhou's Tianhe district. The design is oriented to be luxurious classical European style. This project highlights the aristocracy and magnificence of the sales department, creating unusual visual experiences for the guests. This sales department is divided into two parts according to functional divisions. The front hall is mainly used as reception hall and sand table presentation area, while the hall behind is the negotiation and leisure zone.

The negotiation area has screen with delicate iron patterns as the partition, to make the space display delicacy while being transparent as well. Varying geometric stone patterns, visual effects with softened lines, accompanied with elegant furniture and decorative objects, and the integration of European aristocracy and modern comfort, the collision of classical nicety and modern hardness... All these display the gorgeous and elegant temperament of the sales department.

Jiule Club
九乐会所

设计单位：上海唐玛空间设计有限公司	Design Unit: Shanghai Tangma Space Design Co.,Ltd.
设 计 师：施旭东	Consultants Institution
项目面积：1000 m²	Designer: Shi Xudong
项目地点：安徽省合肥市	Project Area: 1000 m²
撰　　文：施旭东	Project Location: Hefei in Anhui Province
摄 影 师：吴远长	Author: Shi Xudong
	Photographer: Wu Yuanchang

位于合肥的九乐会所周边环境优雅宁静，会所彰显尊贵、私密的氛围。简约的现代时尚感与东方元素的抽象剥离深植于整个空间中，蕴含了大气深邃的东方意境，并具当下的审美形式。设计师从东方传统元素中汲取灵感，大胆加以"破坏"和"否定"，从而创造出一个全新的理念，一个具有时代气息的理念。从对传统东方元素的观察中……在残破的被剥离的传统符号中抽象表达，寻找出最性感的地带，文化、艺术，黑色、白色的表象背后，肆意地表达一种艺术的力量，无形中增强了空间的神秘张力。中国传统磨漆画艺术，浓烈的红色大漆，饱满深沉而斑驳，肌理感含蓄内敛。传统紫砂工艺的提梁壶被大胆"破坏"，以当下的设计形式演变成既具功能性的接待前台又如大厅中的纯白色当代装置艺术品。一颗颗珍珠透过玻璃和光影的洗礼，有序地拉出优美的弧。摇曳的烛光、烛台、朴拙的石材栓马桩、金属装饰品，以及青花瓷的图案，设计师利用色彩、材质、光影以及造型的穿插、对比、和谐所产生的张力，来引起来者的共鸣，传达出空间要述说的东方内在精神。

设计师把接待厅设在会所的高层（整幢楼的腰部），会员从电梯出来，透过充满力量感的几何造型大理石柱之间的空隙，看到楼外的景观，从那一刻起，一段充满诗意的心灵之旅即将展开。黑金龙大理石地面，黑灰色如流水般的纹理从电梯间开始延伸，光洁的理石质地与柔顺的流水纹路碰撞出双重的美感。在这里，中国四大发明的活字印刷被现代材质重新演绎，"道德经"的雕刻文字被设计师平面排序，造型夸张，以及强调尺度感的中式椅子，用亚克力来表现，似乎在告诉来这里的会员完全不一样的中国元素的解读方式。

在全场黑色金属异形天花的映衬下，一如设计师施旭东一贯的风格：强调新东方精神，强调艺术与空间的碰撞。于是在这个充满想象的空间里，精神和意境、品质与灵魂，当代艺术和传统文化开始了邂逅。生命在空间里充盈灵动，从而拥有一份浪漫主义的气质，这个营造了东方文化的艺术空间会成为人们心灵皈依的地带。

This club is located in an elegant and tranquil environment. This club manifests aristocratic and private atmosphere. Concise modern feel together with abstract oriental elements are deeply rooted in the whole space. This profound and grandiose oriental environment embodies the current aesthetic forms. This design draws inspiration from traditional oriental elements, and courageously makes "destructions" and "negations" towards them, thus creating brand-new concepts and concepts with time's breath. The designer tries to get some abstract conveying through observation of traditional oriental elements and the peeled off traditional symbols, to find the perfect zone. Behind the presentation of culture, art, black and white, the strength of art is displayed totally, imperceptibly enhancing the mysterious tension of the space. Traditional Chinese lacquer paintings display intensive red lacquer, appearing profound, implicit and reserved. Loop-handled teapots with traditional red porcelain techniques are courageously destroyed and are transformed into functional reception desks which are like the pure white modern artistic objects in the lobby. The pearls create some elegant arcs in some orderly way through baptism of glass, light and shadow. There are dancing candlelight, candlestick, metal decorative objects and blue and white porcelain. The designer makes use of colors, materials, light and shadow and the tension produced by alternation, contrast and harmonization of formats to arouse the visitors' consonance to display the oriental inner spirits that the space tries to convey.

The designer arranges the reception hall at the high level of the club (which is the waist part of the whole building). After the visitors come out of the lift, through the inter-space between the geometric marble stone pillars with strong sense of power, they can observe the landscape outside the building. From that moment on, a poetic journey for the heart unfold. The black and grey texture like flowing water spreads out from the lift well. Bright and clean marble texture contrasts with flowing water to produce dual sense of beauty. Movable-type printing of China's four great inventions was reinterpreted with modern materials. The carving characters of Tao Te Ching were ordered in a plane by the designer, with exaggerated formats. The Chinese style chairs stressing scale were presented with acrylics. It seems like it is telling people coming here some totally different interpretation methods of Chinese elements.

These traditional oriental culture and art are not simply simple and enumerated replication of symbols, but making public current aesthetic temperament through modern design forms and languages behind the presentations. Set off by black metal ceiling, all displays the consistent style of the designer Shi Xudong, emphasizing on new oriental spirit and the contrast between art and space. In this space full of imaginations, spirit and artistic conceptions, quality and soul, modern art and traditional culture begin these encountering. Life is rich and dynamic in the space, thus possessing some romantic temperament. This artistic space creating oriental culture would become a zone converting to Buddhism or some other religion for the heart.

三省园高尔夫会所

Sansheng Garden Golf Club

设计单位：北京大木博维建筑装饰设计有限公司
设 计 师：张楗波
项目地点：河北省固安县
项目面积：720 m²

Design Unit: Beijing Damu Bov Architectural Decoration and Design Co., Ltd.
Designer: Zhang Jianbo
Project Location: Gu'an in Hebei Province
Project Area: 720 m²

Neo-Classical Reflection·Top Leisure Business Clubs

会所室内设计以自然清新、舒适雅致的北欧风格为主导。在设计上考虑建筑与室内空间的整体性，因此室内空间材料的运用延续了建筑对木材的使用，深色的鸡翅木纹理配以米黄石材，舒适而富有亲和力。

在软装配饰上，贵宾区运用了具有历史性的高尔夫主题老照片及饰品，将空间以叙述的方式串联起来。白色调的北欧风格家具，深色厚重的牛皮拼花地毯，再以朴实无华的陶罐及极具现代感的不锈钢雕塑做点

缀，使得空间氛围朴实，有历史、有文化感和现代感，在装饰性上更整体，与建筑物更为融合。

会所以自然光为主，点光源为辅。白天大面积的落地窗为室内提供充足的自然光；到了夜晚，以暖黄色点光源为主，通过点光源对墙体艺术画的处理来强调空间的层次感觉，营造与白天不同的视觉感受。

舒适高雅，低调奢华，这正是我们在本设计中努力寻找的空间精神气质。

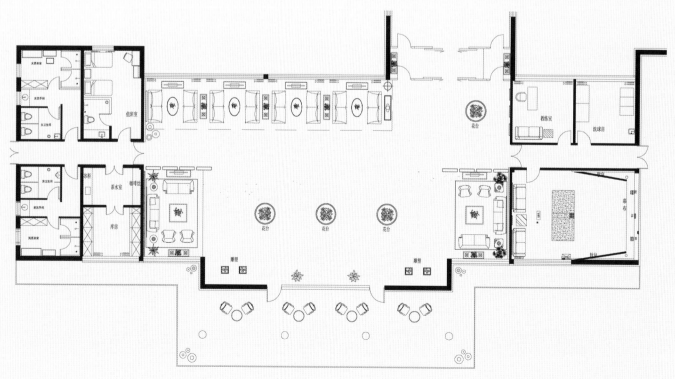

The interior design is centered on natural, fresh, comfortable and elegant north European style. The designer focuses on the wholeness of the architecture with interior space. Thus the application of interior space materials continues the application of wood of the architecture. The combination of dark-color wood grain and beige stone appears comfortable and cordial.

As for the soft decoration ornaments, the VIP area applies the historical pictures and ornaments of Golf theme which connects the space in a narrative way. White north European style furniture, dark color cowhide pattern carpet, primitive pot clay and the ornaments of modern stainless steel sculpture make the space show the primitive, historical, cultural and modern atmosphere which integrates well with the architecture.

This club has natural light as the tone color, and spot light source as the subsidiary color. During the daytime, large area French window provides sufficient natural light for the interior space. Till the night, with warm yellow point light source at the center, through the treatment towards the artistic painting on the wall, the sense of layers of the space is highlighted, to create some different visual sensation from the daytime.

Comfort and elegance, low-key style and luxury, all these are the space spiritual temperament we are pursuing in this design.

虎门名店私人会所
Humen Famous Shop Private Club

设计单位：KSL 设计事务所
设 计 师：林冠成、温旭武、马海泽
项目地点：广东省东莞市
项目面积：1500 m²
主要材料：橡木、灰木纹大理石、黑麻石、皮革、壁纸、黑钢、夹丝玻璃

Design Unit: KSL Design(HK) LTD.
Associate Designers: Andy Lam, Wen Xuwu, Ma Huize
Project Location: Dongguan in Guangdong Province
Project Area: 1500 m²
Major Materials: Oak, Grey Wood Grain Marble, Black Stone, Leather, Wallpaper, Black Steel, Wired Glass

Neo-Classical Reflection·Top Leisure Business Clubs

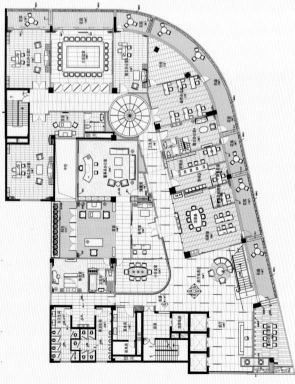

KSL设计事务所对材料及家具极具苛刻的挑选使得本案的室内环境完美地融合在一起，突显出有别于其他设计的高端时尚感。设计师遵循"大繁至简"的设计理念，注重材质给人带来的不同心理感受，而不是仅仅追求外在的形式感。在设计中，将中式符号有意识地化繁为简，再层层深入地渗透到整个空间里，使得空间大气却又不失精致感。

The rigorous selection towards materials and furniture makes the interior environment integrate well, displaying some high-end fashion feel different from other designs. The designer follows the design concept of simple the best, emphasizes on the different psychological feelings that materials leave people with, rather than focus on the exterior forms only. In design, the designer consciously change complicated Chinese symbols into simple ones, then infiltrates them into the whole space layers by layers, to make the space magnificent but delicate as well.

The Reception Club Of Sanya Peninsula Blue Bay

三亚半岛蓝湾接待会所

设计单位：大勺国际设计中心
开 发 商：三亚润丰建设投资有限公司
设 计 师：林宪政、张三巧
软装设计：上海太舍馆贸易有限公司
项目面积：939 m²
主要材料：白色外墙漆、黑色陶砖、透光玉石、地毯

Design Unit: Symmetry Design Center
Developer: Sanya Runfeng Construction and Investment Co., Ltd.
Designer: Xianzhen Lin, Zhang Sanqiao
Soft Decoration design : Shanghai MoGA Decoration Design Company
Project Area: 939 m²
Major Materials: White Exterior Wall Paint, Black Earthenware Brick, Jade Pervious to Light, Carpet

碧海蓝天，心灵无限旷达

半岛蓝湾依托榆林内海湾而建，三面环山，围合于凤凰岭、六道岭、孟果岭等连绵起伏的群山之间。远望中的接待会所建筑，就像是一艘撑起白帆在海中遨游的船，外立面由白色张拉膜撑起作桅杆，端部配以深邃的碗状水体，迎合木百叶制造出的亲水平台，犹如船头舢板，承接出海天一色的壮景，即刻邀您远离城市燥热，感受大自然的恩赐。建筑物外部大量飘檐的运用，能充足遮阳，真正从物理与空间关系的结合出发，达到"自然阴凉、减少空调能耗"的目标。

自然映射，气韵传神

清晨，呼吸一抹清新空气，体味原汁原味的海滨生活；夜晚，建筑照明勾勒出会所的整体外廓。波浪形的木格栅、巨型碗状的水池，整体壁板采用白色线板设计，与室内玉石主题墙发散开的纹理光线相呼应，顶部配以香槟色金属吊灯，营造出梦幻氛围，而黑色小瓷砖铺成的"人"字形地面，一展自然手工的质朴性，与华丽的主题墙形成反差式的视觉对话。

精致空间，悠然开阔

在室内设计中，一楼布置为影音室、模拟样板房；二楼为开放式贵宾洽谈区、天井式中廊将喧闹的销售区隔开，而建筑最深处的玉石背景前，缀着银箔的大吊灯，为整个吧台区营造出游艇般华丽气氛却又不失休闲之味。拾阶而上，行至高点是深邃的水池，转入中廊后，左右两侧分别是独立的开放式 VIP 区和洽谈区，再往后，则由楼梯向下进入多媒体影音室、整体规划介绍区和精装修样板间展示区。

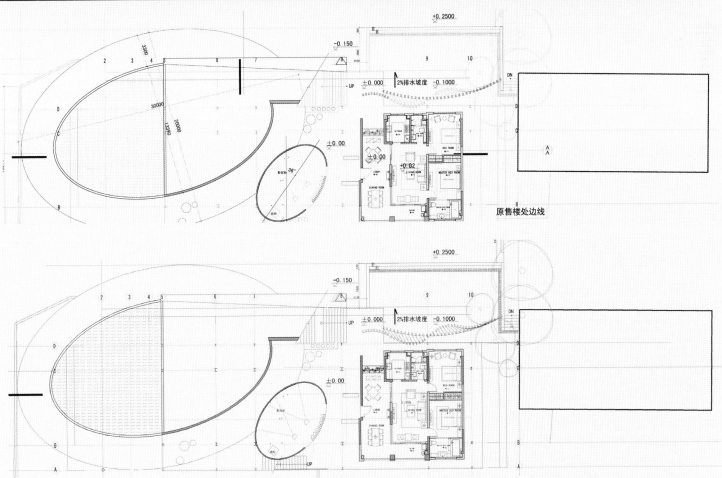

原售楼处边线

With blue sea and sky, you can enjoy the most open and clear heart
Peninsular Blue Bay is built along the inner bay of Yulin, with mountains such as Phoenix Mountain, Liudao Mountain and Mengguo Mountain on three sides. The building of the reception club in the distance is like a sailing boat. The exterior has white membrane structure as the mast. The end is accompanied with bowl-like water corresponding with waterside platform made of wood shutter which is like sampan at the end of the boat to create some spectacular views with sea integrating into the sky. Thus you can get far away from the hustle and bustle of cities and experience the gifts of nature. The eaves of the exterior of the building are designed to keep out sunshine and attain the purpose of "creating natural shade and decreasing energy consumption of air-conditioners" starting from the combination of physical and space relationships.

Reflected by nature and enjoy penetrating charms
In the morning, you can breathe some fresh air and taste the original coastal life here. During the night, the whole outline is displayed by the lighting of the architecture. There are undulating wood gratings and huge bowl-like pool. The board is designed with white line board and corresponds with the pattern rays of light emitting from the interior jade theme wall. The ceiling is designed with champagne metal chandeliers, forming some dreamy atmosphere. Herringbone ground is paved with black ceramic tiles, displaying the simplicity of natural handwork, producing some contrasting visual dialogue with the magnificent theme wall.

The delicate space is carefree and expansive.
As for the interior space design, the first floor is designed to be audio-visual room and model house. The second floor is open negotiation zone for distinguished guests.

The courtyard-like corridor separates it from the noisy sales area. The end of the space is a jade background dotted with silver foil chandeliers, bringing some magnificent but leisurely taste to the bar counter area. While you walk upwards to the highest of the space you can find a water pool. When you turn in to the corridor, you can find independent open VIP zones and sales negotiation area on both sides, when you walk farther, you can go downstairs into the multimedia audio-visual room, planning introduction area and decorated sample house show room.

Aristocrat Club – Red Wine, Cigar and Coffee Bar

贵族会馆——红酒、雪茄、咖啡吧

设计单位：豪思环境艺术顾问设计公司
设 计 师：王严民
项目地点：黑龙江省佳木斯市
项目面积：325 m²
主要材料：锈板瓷砖、文化砖、皮革、浮雕曲柳贴面板、布艺、壁纸、乳胶漆
摄 影 师：王严民

Design Unit: House Environmental Art Consulting Design Co., Ltd.
Designer: Wang Yanmin
Project Location: Kiamusze in Heilongjiang Province
Project Area: 325 m²
Major Materials: Relief Ashtree Veneer, Fabrics, Wallpaper, Emulsion Paint

红酒、雪茄、咖啡吧，隶属贵族会馆的配套项目，华丽高雅，专为私属的VIP会员专享。一进入室内空间，就被一排排木质酒架吸引住了视线，琳琅满目的红酒瓶静静地伫立在酒架上，等待着客人的挑选。设计师在酒架的对面设置了就餐区，黑色皮革的沙发搭配洁白的餐布，突显出高雅的就餐氛围。

沿着台阶缓缓而上，就到了位于二楼的会客区，柔和的灯光让人即刻就沉浸在华贵、典雅的氛围中，暖色调的木质吊顶、璀璨夺目的水晶灯具、墙壁上色彩饱满的油画，无一不显示出设计师对装饰材料极其娴熟的搭配，展现出会所空间的奢华大气。设计师在空间上的巧妙用心，意图在红酒、雪茄、咖啡的环境中增添文化的特质，通过对材质、灯光、装饰品的把握，表现出属于这里的内敛又颇具内涵的高贵气质，给客人以尊贵的享受。

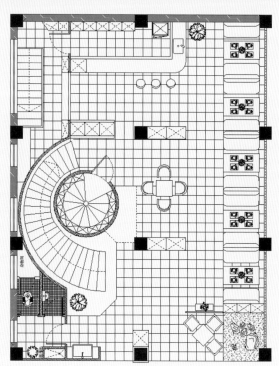

Red wine, cigar and coffee bar is the accessory project for aristocrat club. The whole environment is resplendent and elegant, which is exclusive for VIP members. Upon entering the interior space, your eyesight would be attracted by rank upon rank of wood wine racks. Nice red wine bottles stand quietly on the wine racks, waiting for the guests to pick them up. The designer sets dining area opposite the wine rack. Black leather sofa is matched with white table linen to highlight elegant dining atmosphere.

While you walk upward slowly along the stairs, you can come to the meeting area on the second floor. The soft lighting instantly makes people immerge in the noble and elegant atmosphere. Wood ceiling with white color tone, dazzling crystal lights and the paintings with rich colors on the wall all display the designer's adept collocation towards decoration materials and the magnificent atmosphere of the club space. The designer's ingenious intention towards space adds some cultural quality towards the environment with red wine, cigar and coffee. Through presentation of materials, lights and decorative objects, the designers display the low-key noble temperament with profound connotation, endowing the guests with sublime enjoyment.

碧海国际水疗会所

Bihai International Hydrotherapy Club

设计单位：深圳海外装饰工程有限公司
设 计 师：万文拓
项目地点：河南省郑州市
项目面积：9000 m²

Design Unit: Shenzhen Haiwai Decoration Engineering Co., Ltd.
Designer: Wan Wentuo
Project Location: Zhengzhou in Henan Province
Project Area: 9000 m²

Neo-Classical Reflection·Top Leisure Business Clubs

郑州商都碧海国际水疗会所坐落于河南省郑州市，按五星级标准设计装修，集桑拿、水疗、淋足、按摩、棋牌、客房、桌球室、电影院于一体，是在业界中首次导入健康水疗新酒店理念的休闲会所。

在当代多元化、个性化元素不断注入生活领域的今天，原本传统化、雷同化、单一化的很多社会场所逐渐演变成功能齐全、个性鲜明、设备完善的新兴场所。洗浴业正是其中之一。它不再停留在单纯的清洁身体的基本需求上，休闲娱乐、商务洽谈、酒店餐饮等综合性功能已体现在当下的洗浴场所中。

整个设计融入了土耳其风情，将"奢华的休闲空间"结合桑拿发源地文化，通过对材料、造型、符号的整理和融合得以体现。乳白的基色配以不同颜色的灯光，把空间渲染成异域的感觉：梦幻的居室、明亮的大堂、华贵的休息厅、充满幻想的浴池。镜面、大理石、陶瓷锦砖、花格镂空元素的文化符号及特质石材的运用，巧妙地塑造出空间细腻、恢弘的质感。异域造型符号及纹饰的运用，为空间注入了灵魂，为本案增添了神秘的色彩。

This project is located in Zhengzhou of Henan province. It was designed by a famous designer in Shenzhen whose name is Wan Wentuo. It was designed according to 5-star standard, integrating sauna, hydrotherapy, lavipeditum, massage, chessboard, guest room, table tennis room and cinema. This is the first leisure club in the industry which introduces healthy hydrotherapy new hotel concepts.

In today's world with diversified and personal elements instilling into the life sphere. The original traditional, similar

and simple social locations gradually evolve into new sites with complete functions, distinct characteristics and well-equipped equipments. The bath industry is among these. It not only stays on the pure basic requirements of cleaning the body, many other comprehensive functions such as leisure and entertainment, business negotiation, dining and restaurant all are displayed in the current bath rooms.

The whole design integrates exotic Turkey amorous feelings, integrating luxurious leisure space with the culture of the birthplace of sauna, displayed through the arrangement of materials, symbols and patterns. Accompanied with lights of different colors, the milky tone color portrayed the whole space with some exotic feeling: the dreamy rooms, bright lobbies, noble leisure room and the bathing pool full of illusions. The magnificent and delicate texture of the space was created through cultural symbols and special stones such as mirror, marble, mosaic, lattice hollow-out elements, etc. The application of exotic pattern symbols and graphics instill the space with souls, creating some mysterious colors in the space.

Neo-Classical Reflection·Top Leisure Business Clubs

333

一湖会所
Yihu Club

设 计 师：周少瑜	Designer: Zhou Shaoyu
项目面积：350 m²	Project Area: 350m²
主要材料：青砖、玻璃、钢构、金刚板、壁纸、细木工板等	Major Materials: Blue Black, Glass, Steel Structure, Laminate Flooring, Wallpaper, Blockboard
摄 影 师：唐辉	Photographer: Tang Hui

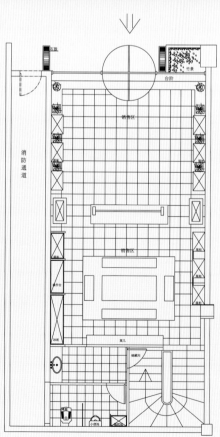

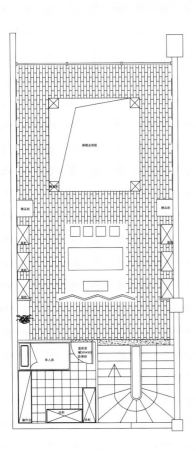

本会所的平面方案设计引用了中式建筑中的"回"及"井"字的构成作为设计元素,采用了中轴线对称的布局形式,营造出整体而又线条分明的格局。在装饰上,没有刻意地使用繁复的造型及材料,而是以简洁的材质和明显的色调对比来刻画空间,使之突显沉稳、大气。在茶室中,设计师采用了中国传统的青花瓷元素来贯穿各个功能空间,让空间连贯、统一。而色彩上采用了中国传统的青、白、灰来表现茶文化的意境。

会所在功能上融入了茶、瓷器、字画、根雕、古玩等艺术品的品鉴及销售功能,也结合了休闲、会谈、会议等商务功能,力求营造出一个具有东方文化底蕴,又兼具现代、时尚的商务文化会所。

The plane design of this club introduces the constitution of traditional Chinese architecture with layout like Chinese characters "回" and "井" as design elements, applying layout formation of axial symmetry to create some complete pattern with distinctive patterns. The decoration does not deliberately use complicated patterns or materials, but depicts the space with concise materials and distinct color tones to make it appear sedate and magnificent. Inside the tea room, the designer applies traditional Chinese blue and white porcelain elements throughout every functional space to make the space coherent and complete. As for colors, the designer applies blue, white and grey of traditional Chinese culture to present the artistic conception of tea culture.

As for function, this club integrates appreciation and sales of artistic objects such as tea, chinaware, calligraphy and painting, root carving and antique, etc. It also combines business functions such as leisure, negotiation, conference, etc., thus to create a business and cultural club with oriental cultural connotations and modern and fashion tastes.

图书在版编目(CIP)数据

映象新古典·顶级休闲商务会所/ ID BOOK工作室编.—武汉：华中科技大学出版社，2013.5
ISBN 978-7-5609-8715-6

Ⅰ．①映… Ⅱ．①I… Ⅲ．①休闲娱乐－服务建筑－室内装饰设计－中国－图集 Ⅳ.①TU247-64

中国版本图书馆CIP数据核字(2013)第030470号

映象新古典·顶级休闲商务会所

ID BOOK工作室 编

出版发行：华中科技大学出版社（中国·武汉）
地　　址：武汉市武昌珞喻路1037号（邮编：430074）
出 版 人：阮海洪

责任编辑：曾　晟　　　　　　　　　　　　　　　　　　　　　　　责任监印：秦　英
责任校对：赵爱华　　　　　　　　　　　　　　　　　　　　　　　装帧设计：张　艳

印　　刷：北京佳信达欣艺术印刷有限公司
开　　本：965 mm×1270 mm　1/16
印　　张：21.5
字　　数：168千字
版　　次：2013年5月第1版　第2次印刷
定　　价：349.00元(USD:86.99)

投稿热线：(010)64155588-8000 hzjztg@163.com
本书若有印装质量问题，请向出版社营销中心调换
全国免费服务热线：400-6679-118 竭诚为您服务
版权所有　侵权必究